Influencing Organ   ology   Thorn

Warwick Thorn

Influencing Organisational Culture through Technology

Developing and Implementing a User-centred Intranet, with Management Interventions, to Encourage a Participative Work Culture

VDM Verlag Dr. Müller

Imprint

Bibliographic information by the German National Library: The German National Library lists this publication at the German National Bibliography; detailed bibliographic information is available on the Internet at http://dnb.d-nb.de.

Any brand names and product names mentioned in this book are subject to trademark, brand or patent protection and are trademarks or registered trademarks of their respective holders. The use of brand names, product names, common names, trade names, product descriptions etc. even without a particular marking in this works is in no way to be construed to mean that such names may be regarded as unrestricted in respect of trademark and brand protection legislation and could thus be used by anyone.

Cover image: www.purestockx.com

Publisher:
VDM Verlag Dr. Müller Aktiengesellschaft & Co. KG, Dudweiler Landstr. 125 a, 66123 Saarbrücken, Germany,
Phone +49 681 9100-698, Fax +49 681 9100-988,
Email: info@vdm-verlag.de

Copyright © 2008 VDM Verlag Dr. Müller Aktiengesellschaft & Co. KG and licensors
All rights reserved. Saarbrücken 2008

Produced in USA and UK by:
Lightning Source Inc., La Vergne, Tennessee, USA
Lightning Source UK Ltd., Milton Keynes, UK
BookSurge LLC, 5341 Dorchester Road, Suite 16, North Charleston, SC 29418, USA

ISBN: 978-3-8364-8681-1

Acknowledgment

The project involved significant leeway in my role as Principal and Development Manager of the organisation where the project was based. I would like to thank Timothy Cooper, my employer and the owner and CEO for his support of the intranet development process and the parallel management interventions involved. I would also like to express my appreciation to my colleagues who have willingly contributed their involvement and ideas - in particular, Vullnet Abdyli the New Media Developer, who assisted me with the technology side of the project. Also, Chris Dowie the Marketing Manager, Charlotte Reynolds and Amy Cooper my personal assistants, and Johannes Balzer, who assisted me with quality assurance management.

I could never have completed the project thesis without your support and assistance.

Behind the intranet project was a thesis as part of my MA in Communication Studies with the Auckland University of Technology. I would also like to thank Dr. Frances Nelson for her patience in guiding me as I struggled with addressing the complex dimensions of organisational culture and her patience as I grappled with how to structure what was to be an evolving discovery. I would also like to acknowledge Dr. Alan Cocker, who has supported me in the course of my study.

Warwick Thorn
June, 2008

Table of Contents

Acknowledgement	1
Table of Contents	2
Lists of Figures	3
Introduction	4-11
SECTION ONE: DEVELOPING AND LAUNCHING THE INTRANET	
1.1 Introduction	12
1.2 The first stage: Prototype development	15
1.3 The second stage: Usability testing of prototype	36
1.4 The third stage: Prototype testing with key users	43
1.5 Taped diary of periodic feedback	47
1.6 The fourth stage: Launching and training for all users	48
1.7 Conclusion	70
SECTION TWO: REVAMPING THE INTRANET	
2.1 Introduction	72
2.2 Revamped intranet: Interface and features	72
2.3 Revamped intranet: Views of staff	85
2.4 Conclusion	108
SECTION THREE: THE DEGREE OF INFLUENCE	
3.1 Introduction	111
3.2 The organisation in a knowledge based and global economy	114
3.3 The effect of a user-centred intranet on organisational communication	118
3.4 The effect of a user-centred intranet on organisational culture	123
3.5 The effect of a user-centred intranet on knowledge management	125
3.6 Contribution of this project to knowledge	127
3.7 Conclusion	130
References	131
Index	136

Lists of Figures

SECTION ONE: DEVELOPING AND LAUNCHING THE INTRANET

Figure 1.2.1 Screenshot of the intranet after the Usability Testing Session	16
Figure 1.2.2 The Silverorange Intranet Homepage	19
Figure 1.2.3 Prototype menu	24
Figure 1.2.4 Project management plan	25
Figure 1.2.5 Screenshot of the Student Welfare data entry reporting page	30
Figure 1.2.6 Screenshot of the Student Welfare reporting page with entries	31
Figure 1.2.7 Screenshot of the ESOL department data entry reporting page	32
Figure 1.2.8 Screenshot of the ESOL department reporting page	33
Figure 1.3.1 Summary notes of the Usability testing sessions	38-41
Figure 1.6.1 Login screen	51
Figure 1.6.2 Screenshot May 2003	52
Figure 1.6.3 Screenshot of the minimised screen	52
Figure 1.6.4 Screenshot of the intranet September 2002	54
Figure 1.6.5 Screenshot of the Academic Allocation report by January 2003	57
Figure 1.6.6 Screenshot of the Contacts page	60
Figure 1.6.7 Screenshot of the Files page	67

SECTION TWO: REVAMPING THE INTRANET

Figure 2.2.1 Login, which does not show again if the user ticks the 'Remember Me' box	74
Figure 2.2.2 The main window	74
Figure 2.2.3 The 'Add News' form	75
Figure 2.2.4 The 'Fix it' form	76
Figure 2.2.5 The 'Fix it' page	77
Figure 2.2.6 The 'Add Message' form for the President, Principal and Marketing manager	78
Figure 2.2.7 The 'Add Contact' form	79
Figure 2.2.8 The 'Forum' page	80
Figure 2.2.9 The update by 2005	82
Figure 2.2.9 The update by 2005 (Cont...)	83
Figure 2.2.9 The update by 2005 (Cont...)	84

SECTION THREE: THE DEGREE OF INFLUENCE

3.3.1 Additional features post revamp stage.	119

Introduction

This story is about the effect on an organisation of developing and implementing a user-centred intranet in an organisation. The intranet was meant as a tool for bringing about wider changes in the organisation - changes to make the organisation more participative. A tool to align internal communications, reporting structures, and how knowledge was managed. The idea was to bring about cultural change in the organisation – to develop a more participative culture where employees were more involved as contributors while having to deal with a high rate of change.

A prototype intranet was developed based on the aim of developing a culture of participation. The prototype was then introduced to key managers through usability testing sessions, and testing with key staff. During this phase, there were a number of iterative technological developments. The intranet was then launched by setting it up on all computers in the organisation, it was introduced to staff. This was supplemented with training. A help file was also built into the intranet itself. All the while, engagement with the organisation meant that initiatives were taken in terms of managing the organisational culture, communication and knowledge management systems. It was an evolutionary endeavour.

> *The intranet was developed ... All the while, management engagement with the organisation meant that initiatives were taken in terms of managing the organisational culture, communication and knowledge management systems.*

The term 'user-centred' implies that users, meaning all staff in the organisation would be contributing rather than being recipients of disseminated information from management.

The organisation was an educational institution and my role in it was as Principal.

I was involved in two ways. Firstly, in a technical development role and secondly in a management role. Technically, I have a background in software development and with assistance from our new media developer and IT staff member we were able to manage the development in-houses. As the Principal I was in a pivotal position to apply supportive and strategic changes across the organisation as necessary. The CEO and owner of the company did have the dominant presence in the organisation, so one question was, to what degree could I really change the culture of the organisation.

> *The term 'user-centred' implies that users, meaning all staff in the organisation would be contributing rather than being recipients of disseminated information from management.*

The integration of organisational activity and technological provisions is a characteristic of our modern technological society. A user-centred intranet, as a mechanism, should coexist as part of an organisations culture with other forms of internal communication and knowledge management systems.

Just as technology in the workplace needs to integrate with other aspects of an organisation, so other aspects of the organisation need to integrate with technology. In order for a user-centred intranet to work effectively the aspects of the organisation that might be made more effective by its implementation will need to be adapted in order to integrate with the technology. My role in the organisation provided me with a means to do just that.

There are various aspects that are reflected on in this book. From a management perspective, the big picture is provided from the change management perspective, notably the interrelatedness of the developments. From a Communications manager's position, the developments apply to internal communications and the place of technological innovation. From the New Media developers perspective the technology questions are addressed as well as how to process a large development in an organisation. From the IT perspective the implications on data security and implementation of a new technology are considered.

What I wanted to explore

There are two things I wanted to prove with this project. One is that a 'user-centred' emphasis in intranet design is necessary in today's working world because more participation is needed than before.

> *There are two things I wanted to prove with this project. One is that a 'user-centred' emphasis in intranet design is necessary in today's working world because more participation is needed than before.*

A post-industrial, knowledge based economy, sometimes referred to as a knowledge-economy, is more decentralised because it needs to be more flexible and adaptive to change. In such an environment, the know-how that workers gain though experience and creative thinking become a valuable asset.

There are embodied world and virtual world ways in which this 'know-how' can be captured.

Another thing I wanted to prove, is that the process of developing a "user-centred" would, in itself, affect a cultural change in the organisation – with communication, and knowledge management in. It can be a tool for changing organisational culture.

> *It can be a tool for changing organisational culture.*

D'Aprix (1999) thinks that ineffective communication processes are a negative product of the beliefs of an organisation's leadership, mirrored in the communication system and behaviour of the organisation. According to D'Aprix, innovation management is needed over creative spaces, networks and virtual networks, to build know-how. I hoped a user-centred intranet would achieve such a purpose.

When I considered these aims I understood that I needed to fully engage with the culture of the organisation. On the one hand, I managed the intranet development and implementation. On the other hand, I managed the parallel change process of the organisation – including changing the organisations reporting structure, channels of communication, and the structure and format of information stored on the network drive.

Why I took on this project

I had a number of reasons to take on this project. I had been interested in the cultural impact and socio-political affects of uses of the Internet. I had also been interested in how reality television as a medium could create a certain type of culture. Could I also create or at least affect our organisational culture though this project. Some years ago I had also been interested in the interactive potential of computers. Similarly the communicative potential of Internet-based technology captured my imagination. I also wanted to provide management leadership towards a more participative working culture. I was in a unique position, in that, I had the opportunity to do this in an organisation with relative autonomy.

I wanted to embrace a networked organisational management structure. This seemed appropriate governing the ever-changing market. I had been very frustrated over how our internal networks sometimes disregarded line management protocols. I saw this very clearly when I created an organic chart showing how staff networked and involved themselves in different areas of the organisation. This was achieved through a staff training day when I got all staff to show by diagram how they communicated with people in the organisation. It was very clear that some people disregarded communicating with their line manager and some only communicated with their line manager. Both these extremes created clear problems in the organisation. I wanted people to communicate, of course with their line mangers and also network in the organisation with those affected by their work. Could this project, governing it was about communicative technology improve this situation? I hoped so.

I wanted to embrace a networked organisational management structure.

... some people disregarded communicating with their line manager and some only communicated with their line manager. Both these extremes created clear problems in the organisation.

While I was working on this project, I was also embroiled in all the pressures of managing an organisation which is in constant change. In theory, I should have been able to implement an intranet easily, but I had to contend with the issues of organisational politics in an entrepreneurial culture and a degree of consequential disorganisation, with growth in a ballooning market and with the bursting of that balloon, and also with departments which sometimes do not work together.

While I was working on this project, I was also embroiled in all the pressures of managing an organisation which is in constant change.

I also had to contend with the conflict which can arise between information-technology (IT) and new media designers. The conflict arises from the traditional mandate IT departments have to control security integrity, and the mandate that new media designers have to work in creative ways. I needed both to work together cooperatively. For example, if there was to be a photo section and the intranet was to be stored on an external server then the costs charged to the organisation by its Internet service provider would increase because of the increased traffic caused by users viewing the pictures. I had to look at these kinds of questions by involving both the IT and New Media staff.

I also had to deal with staff turnover. Allocating a staff member to revamping the organisation's manual was frustrated when they left the organisation before fully completing that project and budget restraints frustrated attempts to reallocate that task. I also had to deal with priority shifts. There were times when I needed to step out of the project in order to attend to completing other tasks which had an urgent priority. This project is then very much an applied project, where the aims are tested in a challenging 'rubber meets the road' environment.

This project is then very much an applied project, where the aims are tested in a challenging 'rubber meets the road' environment.

This project provides insights relevant for communications and new media professionals with whom managers of the modern organisation increasingly interact.

> *This project provides insights relevant for communications and new media professionals with whom managers of the modern organisation increasingly interact.*

Move towards user-centred intranets

An intranet is the same as a website in terms of the technology it is built with. A company website, which provides a sales and marketing interface, often offers customer support. A website is publicly accessible with guest and membership areas. However, public access is prevented to an organisation's intranet. An intranet provides an interface accessible only by staff within an organisation. Public access can be prevented by a top level login screen or it can be prevented by being stored on an organisation's internal network, which technically disconnects it from public access. The context and functionality of an intranet is purpose-built to serve an organisation's internal needs. An intranet may include information stored in a database and access to files within an organisation's network.

Computer access and staff competency also affects whether an intranet can be effective. For staff to be able to access an intranet they need access to a computer. Since not all staff may be able to do this, staff who can access an organisation's intranet may be a subset of the organisation's employees. Some staff may also be less confident when using computers and therefore may need training in order to be able to share participatively through a user-centred intranet. This was certainly evident during this project. If an intranet is stored on the Internet with login access, staff who are geographically dispersed can access it. This can include staff working in divisions in different regions, staff working from home and staff working off site, such as when on a marketing trip. This can be a benefit especially if because of the communicative aim of the intranet that users feel connected to staff in other locations. At one point once the Intranet was up and running I was overseas on a project for the company. I certainly felt good that I could post messages, as on a noticeboard from the other side of the world, knowing that all the staff would see my message on their computers.

Since the early 1990s, intranets have moved from being internal organisational websites which centralised information to mechanisms that enable staff to manage information and interface communicatively with each other in the organisation. The technology of community networks continues to grow.

Intranet design issues have been challenged to embody organisational processes by combining the information processing capacity of information technologies and the creative and innovative capacity of staff, creating a user-centred paradigm which focuses on the participatory potential of intranets. This could represent an organisation's endeavour to draw upon the imaginative and creative potential of an organization's staff. There is a complimentary coexistence between staff participation and collaboration provided by a user-centred intranet.

> *Intranet design issues have been challenged to embody ... [amongst other things] the creative and innovative capacity of staff, creating a user-centred paradigm which focuses on the participatory potential of intranets.*

Book structure

This introductory chapter has explained my role in the organisation and in the project and has introduced the content, reason and intention.

There are blocked sections at the bottom of the pages which contain: an overview of the organisation, theory, and the characteristics of a networked organisational.

The overview of the organisation introduces the key management personnel and strategic partners are introduced. While I was the Principal, I was not the owner. He managed the organisation as CEO on a day to day basis. The impact of the owner's entrepreneurial style is explained, and the story shows how that style affects distinctive cultural features. The organisation's place as an international educational provider in the global marketplace is explained. These aspects contribute to a state of change for the organisation revolving around its response to the marketplace and its restructuring cycles. Internal politics and staffing changes are also introduced, to the extent that they impact on the internal dynamics and thus impact on the aims of the project.

There is theory content that refers to other literature that affects the strategic approaches to technology. The characteristics of a networked organisation are also explored.

In section one a story is told from an insider's point of view. The chapter reflects on the development and implementation stages and the coexistent management engagement.

In section two the development stages are tracked, making a number of discoveries. Implications are drawn in terms of the aim of the project.

In section three conclusions are made so as to inform students of and professionals of communication management and those concerned with networked organisations and the issues of technological integration.

SECTION ONE

DEVELOPING AND LAUNCHING THE INTRANET

1.1 Introduction

In this chapter the project is outlined from the prototype stage to an established intranet. The final revamp of the intranet is the focus of the next section.

There are a number of elements involved in the project. The purpose is to develop and implement a user-centred intranet with the intention of engaging and exploring the cultural, communicative and knowledge management issues. As the Principal of the educational organisation where the project was undertaken with the intention of using the project as an opportunity to try and influence the organisation towards a networked organisational model.

> *The project was undertaken with the intention of ... trying to influence the organisation towards a networked organisational model.*

The overriding goal of the project was to explore the theory of user-centredness, of networked organisations and in particular the need for cohesive participative communication.

I was able to gain access to events and groups that are otherwise inaccessible to external consultants. I had the opportunity to manipulate events to test the boundaries of my theories.

Introduction to the organisation

The following blocked text introduces the organisation and its history and reflects on the manner in which the owner's entrepreneurial style impacts on the culture of the organisation. I also explain my position in the organisation and the relationship of my professional life to the project.

I believe my impact is one of influence rather than a setting of the culture.

I realised the staff in the company were going to have different views about what I was trying to do and about the value of the intranet. In fact I have included some challenging statements from staff. This shows that I needed to hold to a degree of faith in what I was doing. At times the degree of influence I hoped to have did seem rather slim. At the time of writing there have been three years since I left the company – the organisation is still using the intranet.

For the purposes of collecting data for the thesis that underlies this project I followed Schein's (1993, p. 169-170) approach. Schein proposed a "interactive clinical perspective" which involves "series of interviews with individuals and groups of staff, geared to discovering shared underlying assumptions."

Stages in the project development

The first prototype development stage was based on a review of literature on intranet and technology prototyping and implementation.

Overview of the organisation

The organisation which is the host of the intranet development project is Romus Colleges, an educational institution situated in Auckland. Romus Colleges presently provides education for international students in the areas of English for Speakers of Other Languages (ESOL), business studies at the first and second year degree levels, teacher training for ESOL (TESOL) and studies in applied film and television production. Consequently there are four departments in the school.

Over the course of this project the student numbers ranged between 250 and 1,400, because of a boom and bust phenomenon in the market for international students from China.

The college was started as a language school in 1987 by the owner, Thomas Chapel, whose first endeavour in the field of international education was in 1985, when he organised and ran a Japanese tour group for English language tuition and rugby experience. Thomas provided English language tuition and organised for some of the New Zealand All Blacks to coach the students and to play rugby with them. Thomas' entrepreneurial style pervades the school to this day.

The second usability testing stage involved sessions with a number of key staff, who tested and provided feedback on the prototype. This helped with debugging and preliminary improvements to the design. It also helped to gain support from key staff.

The third initial roll out stage involved putting the intranet on the computers of key users. At this stage their trailing of the intranet in their work day informed further improvements and developments of the design. There were numerous iterative changes to the intranet at this time and until the final stage.

> During these three stages I kept an audio taped diary of sporadic feedback. The users would periodically mention issues to do with the intranet. From these issues I would often take small decisions in the evolving design process.

The fourth launch stage involved setting up the intranet on all computers in the organisation and to train staff in its use. Once it had been in use for enough time for staff to form opinions a questionnaire was undertaken to examine opinions on it and how staff generally used their computers, how they accessed files on the network and how they communicated in the organisation (both through technology and face-to-face).

The fifth revamp stage was undertaken once the intranet had significantly evolved and had become accepted within the organisation. At this stage a

Management personnel in the organisation

Two of the longest serving personnel are Charles Dowie, in the position of Marketing Manager, and me as Principal. We have both been with the college from its establishment. We all explore areas for development and growth, and tend to work with enthusiasm when we are picking up ideas and building them into the framework of the college. The internal environment at Romus is often turbulent as we try to keep pace with rapid change.

On top of this entrepreneurial culture we each bring a particular emphasis. Thomas drives the financial impetus and accountability, Charles generates marketing impetus and activity and I emphasise organisational strategy, project management and cohesion. I was responsible for project managing the development and establishment of each of the departments and the educational programmes within them. Thomas does the deals, I make the product and Charles sells it.

number of interactive clinical interviews were undertaken to evaluate the usefulness of the intranet the impact on the culture of the organisation.

1.2 The first stage: Prototype development

The first prototype development stage involved producing an intranet using the user centred paradigm.

One initiative of this time was to think of a way of disseminating company policies, while at the same time enabling user updating of the information. I wrote some software that created an auto-generated menu on the intranet.

Owner and president of the organisation

Thomas Chapel's influence affects the organisational culture most significantly. He is the President of the College. Thomas is an entrepreneur. This means he is often developing new business ideas.

Virtanen (1997) suggests the following multidimensional definition of entrepreneurship with specific emphasis on the entrepreneur as the main actor in the process:

Entrepreneurship is a dynamic process created and managed by an individual (the entrepreneur), which strives to exploit economic innovation to create new value in the market. An entrepreneur is a person, who has entrepreneurial mind with a strong need for achievement. (Virtanen, 1997, p.6)

Entrepreneurs, as Virtanen (1997) says, feel compelled to do their own thing in their own way, needing the freedom to choose and to act according to their own perception. Thomas displays this characteristic of entrepreneurs.

Figure 1.2.1 Screenshot of the intranet after the Usability Testing Sessions

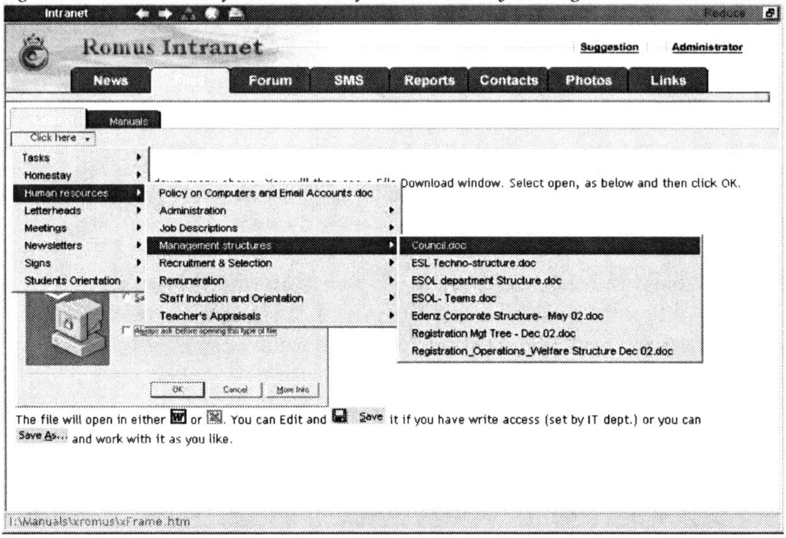

At Romus there is a degree of drama. In entrepreneurial fashion, Thomas is not troubled by the ambiguity and uncertainty that this fast pace of change brings. Particularly during tough business periods, Thomas will involve himself in every aspect of the organization. When resources are scarce, his attention is on what is essential to maintaining viability, which means his attention is not as focused on procedures as much as outcomes. Some staff feel that changes should be processed in a more organized fashion, but that is not the way things are done at Romus.

Romus is organic. Issues of delegating authority are sometimes compromised because staff work across departmental boundaries. Typically, staff are involved in many varied aspects of the organisation. There are established line management protocols, but entrepreneurial culture tends to resist structure and control. Thomas' direct approach leads him to seek information directly from its source, often bypassing the structured chains of authority and responsibility, which can lead to managers feeling undercut, particularly in cases where there is a lack of communication as well. Authority exists more when staff work proactively, accepting unceasing role positioning as a result of organic changes. From this perspective, managers are able to work as an enthusiastic team. There is drama, there is fluid professional structure, and there is vibrancy.

The menus reflects exactly how the files are stored on the shared network drive. The idea was to provide easy and clear access to the files. Under the hood, each night, all these files (*.doc ans *.xls) were auto-converted into html files moved to an internet location. This meant that for users with external Internet access, the menu would auto build as an html version. They could read files, but not edit the originals (because they cannot access the originals on the internal network drive).

The auto-menu would build by searching the drives and files, so if files were deleted or added the menu would recognise it and adjust.

This system to access the files on the shared network drive was a lot of work. Did it get used by the users? No! Was it scrapped? Yes. Basically I was trying to compete with the the standard Windows interface for accessing files. Users were already accessing files that way and external users, basically had hard copied of the files that really mattered to them. That was that. I couldn't change that behaviour and had to accept it. There was some spin-off benefit in that I I had spent a lot of time re-sorting and culling folders and files and tightening the security control of the files on the network drive.

> *This system to access the files on the shared network drive was a lot of work. Did it get used by the users? No!*

Cultural features of the organisation

As an educational organisation our culture includes certain features. It is all about people teaching and learning, and their expectations and the services we sell. It is about student welfare and educational concerns. Within the context of the programmes we run, there is a diverse group of individuals who have often travelled extensively. Within our student body there are also diverse ethnic cultures. This cultural diversity also exists within the staff. The staff is made up of people from Japan, Korea, China, South America, Eastern Europe, Philippines, England, Canada, America, and New Zealand.

Silverorange provided the winning entry of the Neilson Norman Group's intranet Design Annual (2001). I referred to Neilson's (2001) report of winning intranet designs to inform my process. Naturally I wanted to know what was being done elsewhere, and why it was or was not working. Because the report was richly illustrated, it gave me an opportunity to see good intranet designs that are usually hidden behind firewalls. Themes such as letting employees update content, collaboration tools that let employees exchange information through discussion groups, and an emphasis on communication by encouraging departments to post news and other information of interest to different groups, were of note.

Silverorange, provided an interactive demonstration of its intranet on its website. In particular the front page, seen below, offered a news page, for staff to contribute to. I liked this idea and replicated it for our prototype design.

My role in the culture of the organisation

In a sense, I am the owner's right hand man. As the Principal my role is varied. I maintain a facilitative style of line management for the academic departments, facilitating the resourcing for the departments, assisting in curriculum development. As well as that, I also oversee system support, student welfare, assisting in a human resources role, establishing quality assurance. I act as project manager for new developments. I tackle the demands of communication management and instigate strategic initiatives. Organisationally I drive restructuring. In the sense of the game analogy, where the game keeps changing, I work to establish what the rules could or should be and try to provide cohesion for teamwork. As I have the trust of the owner, I can significantly influence how the organisation works. I therefore provide a certain influence and position which means I can undertake various projects, including this thesis project.

Figure 1.2.2 The Silverorange Intranet Homepage.

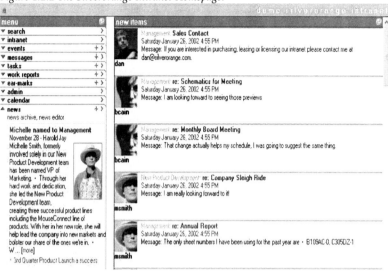

IT and New Media departments of the organisation

Information Technology (IT) management traditionally provides a controlled environment for an organisation's information management. Systems administration determines the hardware and software systems for an organisation shared network drives, mail servers, and Internet access. This requires establishing and maintaining system integrity and security protocols. In 2002 the college's IT was managed by an external consultant who was called upon from time to time, as well as a part time IT student who undertook maintenance tasks. A new educational department manager, who had previous IT systems management experience, began to provide a more strategic approach to our system.

New media management encompasses developing, revamping and maintaining internal and external technologically based organisational media. These include websites, intranets and marketing media such as CD-ROM or DVD. As a discipline area it is unlike IT in that it is driven by creativity. In the case of Romus Colleges Vaughn, the new media specialist principally reported to the Marketing Manager and myself. He also reported to the IT department for matters where website and intranet operation required system protocol settings and as a back up person for IT.

Technology used

To get the first prototype development stage off the ground, I met with the Romus new media designer and we explored the Internet looking for ideas that would apply and be appealing at Romus. We were to draw on my software experience in designing educational language learning software and web page development and on the experience of the new media designer in interactive web page design. One main concern at this stage was to select appropriate technologies to build the intranet.

The Silverorange company was contacted with the idea of purchasing the source code and redesigning modules. The cost for this was prohibitive for this project. Other intranet packages available online were very standard and would not have enabled us to evolve our own model. A design strategy was therefore formulated which would, for the prototype development, involve us using the following software resources:
Authorware Professional (AP), from Macromedia (writer's expertise)
An add-on to AP, called WebFX, from MEDIA shoppe (writer's expertise)
An add-on to AP, called Buddy API, from Magic Modules (writer's expertise)

IT and New Media departments of the organisation (Cont...)

There was not the degree of cohesion between IT and New Media development to collaborate in all matters. This was partly due to the pressures and independent roles each played. The person responsible for IT was also a manager of a department and so had a range of responsibilities in the organisation. He had a very structured approach to IT problem solving. He liked to consider ramifications properly and discuss them with all parties that should have input. Pressures of work made this discussion process difficult at times. There were three IT staff he could allocate work to in the IT area but because they worked part time and also for other managers this was something that added to the frustration of the IT manager. The New Media developer was one who would help out with IT issues, but mainly he reported to the Marketing Manager, myself and the owner, who would also give him independent direction. At a typical meeting between the IT Manager and the New Media developer both would discuss perspectives, but specific agreement was 'left in the air' and rather the priorities of each would affect the outcome.

UltraDev, from Macromedia (New Media Developer's expertise)
Flash, from Macromedia (New Media Developer's expertise)
iTab software (iTab Pro QuickNavBar Type II) from IMINT.com
WordConverter, from SoftInterface, Inc. (writer's expertise)

Authorware Professional (AP) was used to make the frame within which the intranet works and to collect and build information on the organisation's quality manual files. Using an add-on called WebFX, live online content is viewable within this window. Another add-on called Buddy API accesses Windows operation system functionality. Two other tools worked in sync with this platform; WordConverter for external auto-conversion of Word and Excel files to HTML format and iTab for auto-building a quality manual sub-menu interface.

UltraDev was used for developing the online interactive elements, such as the news page and the forum. Flash was used for some aspects of graphic design and menu building.

Free source code to help in building the HTML/ASP based side of the intranet was gleaned from various developer sources on the Net.

The AP software using the Buddy API add-on collected the information on the contents of two folders, where the organisations manual files are kept. It then created a parallel folder and sub-folder structure and sent commands to the WordConverter program to auto-convert all the Word and Excel manual files to HTML copies, storing them into the parallel folder and sub-folder structure. The record of that was stored to a text file. When the intranet was started the text file is read by the iTab software auto-creating an updated menu system for the manuals. This sub-menu needed to be auto-built like this so that it included any new or updated files. By using this approach we were able to explore the options of moving an auto-updating HTML version of the organisations manuals onto the Intranet. While this explains the authoring processes from a technical perspective, the project was evolving and the final revamp culled out the auto-generating menu system for reasons which are explained in this and the following chapter.

Proposal

In June 2002 a proposal was put to the organisation and agreement was given from the owner of the organisation to proceed. The main concerns of the owner were how it would work on the company's network and issues of usage costs. The proposal was set out in a 10 page document with information under the following headings: Introduction, History of intranets, Design considerations, Pertinent points from the market leaders, Issues for getting staff to use an intranet, Other strategic issues, Reflections on the best intranets, Intranet checklist, Project management timeline, Preliminary design, Success measurements, and Acceptance.

Each of these sections was concluded with an agreement scale and a space for comment. The owner was positive about the user-centred approach.

Following the meeting where agreement was given to proceed, I made some entries in the taped diary:

From the taped diary dated 10 July, 2002

> *There was some discussion of how an intranet would work on the company's network. Did it work on each computer or did it use the Internet? What effect would it have on the Internet Service Provider's (ISP) usage costing? There was therefore clarification and discussion on how we would be setting up databases with the organisations ISP. Also that significant increase in usage cost was not expected.*

From the taped diary dated 10 July, 2002

> *There was discussion on the use of the company operational manuals. The company had problems in getting department managers to refer to them and update them as necessary. The owner was interested in whether the Intranet could fix this. It was explained that the Intranet would endeavour to incorporate easy reference to the manuals.*

As a result, the following agreement was added into the proposal, thereby giving me the rights to publish the development story and research data:

> *While completion of the project requires a working copy to be located on the Romus' ISP server, a backup copy of all software code and source files, with documentation will be provided to Romus Colleges on CD-R. The Project Designer will retain copy write of the development story and research data for the purpose of future publication. Any screen shots used for future publication, will have any confidential data extracted prior to publishing. Any future publication would likely refer to the true identity of Romus Colleges and respect professional confidentiality.*

By this stage the intranet prototype development was underway. We had got as far as designing a rudimentary menu. This screen shot, which included notes on the prototype approach, was included in the proposal.

Figure 1.2.3 Prototype menu

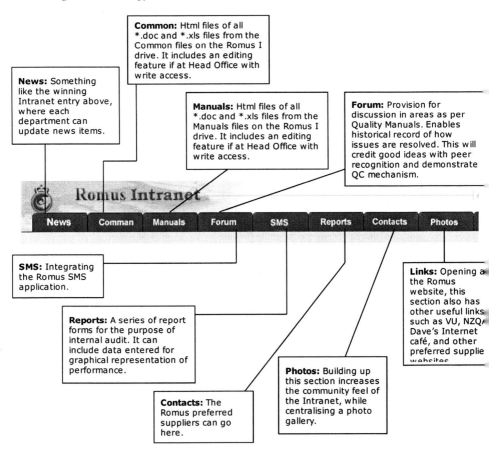

A project management time line was included in the proposal. A tick ☑ has been put next to the items completed by November. The intention was to complete the remaining items over the course of the following 6 -12 months. This was partly achieved but hampered due to work pressures and other work priorities overwhelming the project. My ability to drive the integration process of the intranet usage became refocused again near the beginning of March 2004.

Figure 1.2.4 Project management plan.

June 2002	July	August	September	October	November
Define menu selections and basic design ☑					
Explore ASP code options ☑					
	Apply design considerations ☑				
	Build menu and key task functionality ☑				
		Launch Intranet	Interaction (improve design)	Interaction (improve design)	Interaction (improve design)	Interaction (improve design)	Interaction (improve design)
		Make poll on front page and collect poll data ☑	Remake poll on front page and collect poll data ☑	Remake poll on front page and collect poll data	Remake poll on front page and collect poll data	Remake poll on front page and collect poll data	Remake poll on front page and collect poll data
	Help Heads to think up news items ☑	Training session for heads to enter news items ☑			
		Help Key users to think up improvement ideas ☑	Training session on entering improvement ideas	Meet with key users to show how their feedback has shaped content	
			Help Key users to think up items for forum entry ☑	Training session on using forum ☑	Meet with key users to show how their forum input has been responded to and impacted on work practice	
		User testing for core task (adding news items) ☑	User testing for core task (adding news items) User goal focus	User testing for core task (accessing policy information) User goal focus	User testing for core task (using forum) User goal focus	User testing for core task (entering contact and personal details) User goal focus
			Check and remove impediments for departments to keep their information live and dynamic			
				Apply Intranet checklist		
					Survey management and re-grade proposal agreement scales	Present final report to Management Team Meeting

During the development process there was a checking in with key staff to build buy-in and enthusiasm.

From the taped diary dated 2 August, 2002

> *I have grabbed some key people to have a look at what we were doing. ... The effect of this is to give them the opportunity to ... make a few comments, and sometimes it was just about creating enthusiasm for the idea. ... We would give them a quick demo of parts of what we were developing.*

Another comment, which reiterated this tactic from the taped diary dated 10 July, 2002

> *I suppose as a developer, the question is, you see people and you say to yourself, "They're key", so it's good to show them and get them to buy in, perhaps more than others. You sort of select who you think needs to be enthused.*

It became clear during the development stage that there needed to be a parallel management intervention in the organisation to centralise the way folders were ordered on the shared network drive, and create and maintain some rules for standardisation. From the taped diary dated 8 August, 2002

> *In this company, some people who do that make huge mistakes like duplicating files by doing a copy and then putting a folder in another place, where they think, it is an improvement. What they have done is duplicated files so that two people, or groups of people in the organization are accessing different files and they maybe update them and they become different. This issue, and the issue of people building sub-sub-sub folders because they think that's logical from their own point of view, raises the problem of what standards we have for our file systems. If the organization, as in this one, doesn't have that controlled ... it creates a problem in terms of our intranet ... because we have allocated with agreement with the I.T. Manager, folders - one called common and the other called manuals. This is where the files will be turned into HTML files and the file structure duplicated with the HTML files on the intranet. Therefore a systematic policy on where files are kept and where they are ordered needs to exist. It all affects the benefit of what an*

> intranet can offer. The intranet must offer something that is significantly better than any other way of doing it. So it requires management involvement at this level of the company.

The organisation had been incurring dramatic growth and this was felt the most in the ESOL department where student numbers increased from 200 to 1,000 during this stage of the project. As the organisation grew the communicative culture was changing. The ESOL teachers in particular felt cut off from management and that they did not know what was happening in the company. From the taped diary dated 28 August, 2002

> At the moment the teaching staff are concerned about not getting information and about some holes in the school systems, so whilst the intranet is not up and running it's a good time to present some information to the teachers about what the intranet is going to look like. I took some screen shots just before release, of the news page and some of the forum and made some hand outs, which I took to an ESOL department staff meeting. ... I thought it was one way of satisfying the stress that the staff are under and also getting them on board in terms of the idea ... teachers will be able to put some comments on the forum, which a quality control person can be responsible for checking and can make sure that the issues raised on the forum are drawn to conclusions and built into the systems and policies of the school in a proper way. And secondly, the news page and front page is run in a certain way that people in the organization can get information out to others.

At this time restructuring was being undertaken. I was concerned about how records were being kept and that there needed to be more centralised control of this. I looked to the intranet as being a potential mechanism for this.

> As well as the issue of centralising the way folders are ordered on the shared drive and standardisation rules, intervention was needed to review and update the company's manuals themselves. They had largely fallen out of use. Any such intervention was going to take into account how they might be moved from hard copy to computer accessible files. After attending an external workshop on manual writing I recorded some ideas on this subject. From the taped diary dated 12 August, 2002.

> An intranet is a disseminator of information, and often of company manuals. ... One significant benefit would be to make the quality manuals more available and accessible, through the intranet, and to ensure they are up to date. So therefore, in this case at Romus Colleges, the issue of reviewing the quality of the manuals becomes a parallel management activity of developing the intranet. A number of issues are:
>
> 1. One of the purposes of this intranet can be to encourage staff at the head office to easily access and easily update the original quality manual files.
>
> 2. I will need to review the existing manuals along the lines of how they are structured ... so that when they are viewed on line they become user friendly ...
> 3. The pitch of language is another consideration. Was the language in the manuals pitched to the particular levels of the readers? So that is something we are considering, also simplifying language is logical for content on an intranet.
> 4. The manuals need to be written to incorporate the differences of other Romus' campuses to avoid resentment against head office.
> 5. Another point is made, to never take for granted that other people will know what is in the quality manuals. Therefore one has to consider how that can be taken on as an issue.
> 6. Another point when considering simplifying documentation, especially when we are going to simplify the documentation, is the risk of not documenting any particular procedure.

My management intervention on this was to recruit a staff member to work on it as a project. The strategy was to check all files on the shared network drive for what might be duplicate policies and visit each department to check through these and delete any old policies. The idea was then that we would break up the larger files into files with names that explained the content and enlist department managers to restructure the logic on how the policies, procedures and control documents were filed. We managed to accomplish the first stage of this so that duplicates were reduced to a minimum and we rebuilt the folder structure from which the files could be located. At the same time that we were doing this we tried to inform people, particularly the owner, of the changes that were being made. This project faltered because the staff member working on the project left the company after an initial resorting of the files was accomplished and a culling out of duplicate policy documents had been completed. There was no opportunity to edit the documentation as planned.

Reporting

There were a number of attempts of building departmental reporting systems into the intranet. While these were attempted, in the end they were discarded. The reasons were that were were better off to reconstruct our face to face and paper based reporting because a key issue in reporting is managements response to it – effectively reporting seems most appropriate at a management meeting. While the idea seemed good to me to control reporting via the intranet, it was more important to have effective response and follow up to reporting. My focus on this shifted to reconstitute our reporting paper based forms and our management meeting and I accepted that it was not going to fly as an Intranet initiative.

> *Effectively reporting seems most appropriate at a management meeting.*

Nevertheless a lot of work went into trying to get reporting happen via the intranet.

Introduction to theory

An intranet affects the culture and communication of an organisation. Whether that influence is positive or negative depends on the way the intranet is developed and implemented. It can be imposed as a fait accompli, or negotiated with staff so that it serves their needs and interests.

I believe that a user-centred intranet is strategically important for modern organisations. Technology coexists as part of an organisation's culture because we use technology as one mode of communication and knowledge management. Modern day organisations need the participative and creative involvement of staff because of the fast pace of change which faces modern organisations. Coping with this change is addressed with a user-centred emphasis for the intranet.

I had hoped to get student absenteeism information reported on the intranet. The Student Welfare department followed up on absenteeism and took some initiatives with counselling. Keeping these records online seemed a good idea because from time to time other managers were contacted by agents or parents and it would be useful to be able to access at least summary details via the Intranet.

Figure 1.2.5 Screenshot of the Student Welfare data entry reporting page.

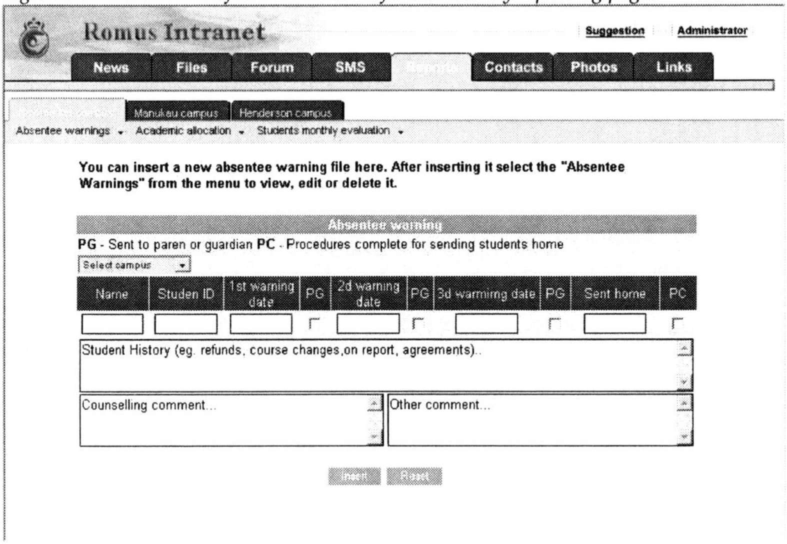

Figure 1.2.6 Screenshot of the Student Welfare reporting page with entries.

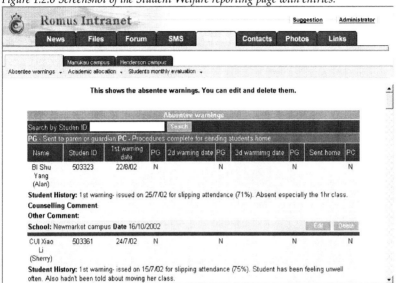

Warnings from the Philosophy of Technology

The German philosopher Heidegger devised a philosophy on the nature of technology, in relation to how people exist in the world. His essay *The Question Concerning Technology* (1949) prescribes a state of being that makes it possible for people to relate to technology in a way that ensures that they remain emancipated. Heidegger (1949) proposes that people need to realise truth exists not in the culture in which we live, but rather in a questioning and thinking attitude – this attitude should be fostered in the way we make technology work for us. That requires participation, hense the user-centred focus of the project.

I had hoped to manage academic department reporting on the intranet as far as numbers by course and management and administrative allocations were concerned. I wanted to do this because management and marketing both had an interest in seeing the trends. I unrealistically hoped hoped to be able to manage all these reporting mechanisms via the intranet.

Figure 1.2.7 Screenshot of the ESOL department data entry reporting page.

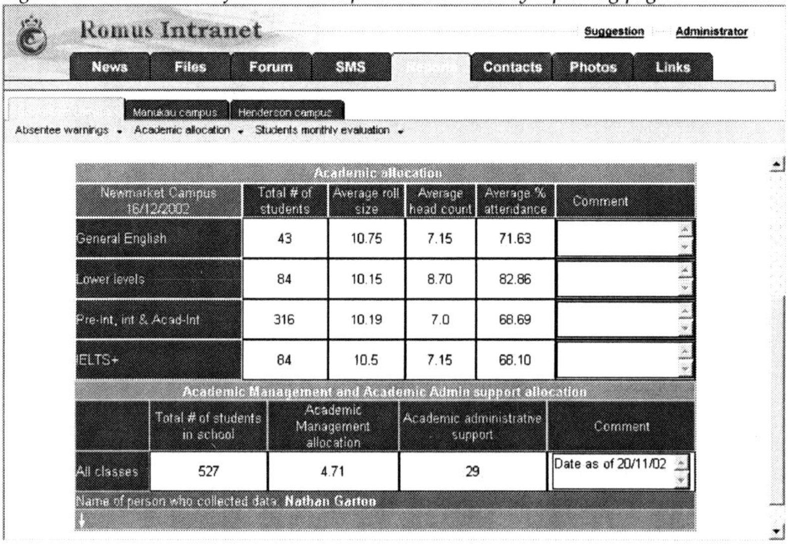
Romus Intranet
Suggestion
Administrator
News
Files
Forum
SMS
Contacts
Photos
Links
Manukau campus
Henderson campus
Absentee warnings
Academic allocation
Students monthly evaluation
Academic allocation
Newmarket Campus 16/12/2002
Total # of students
Average roll size
Average head count
Average % attendance
Comment
General English
43
10.75
7.15
71.63
Lower levels
84
10.15
8.70
82.86
Pre-Int, Int & Acad-Int
316
10.19
7.0
68.69
IELTS+
84
10.5
7.15
68.10
Academic Management and Academic Admin support allocation
Total # of students in school
Academic Management allocation
Academic administrative support
Comment
All classes
527
4.71
29
Date as of 20/11/02
Name of person who collected data: Nathan Garton

Figure 1.2.8 Screenshot of the ESOL department reporting page.

Academic Allocation

General English Total # of Students

Student #	43	47	51	46	61
Date	20Nov02	14/3/03	15/4/03	9/5/03	end 1/7/03

General English Average roll size

Ss per class	10.75	9.4	10.2	9.2	10.17
Date	20Nov02	14/3/03	15/4/03	9/5/03	end 1/7/03

General English Average Attendance

% attendance	71.63%	85.11%	83.53%	76.96%	65.57%
Date	20Nov02	14/3/03	15/4/03	9/5/03	end 1/7/03

Lower Levels, pre-int # of Students

Student #	84	134	88	61	35
Date	20Nov02	14/3/03	15/4/03	9/5/03	end 1/7/03

Lower levels Average roll size

Ss per class	10.15	11.17	11	8.71	7
Date	20Nov02	14/3/03	15/4/03	9/5/03	end 1/7/03

Lower levels Average Attendance

% attendance	82.86%	71.79%	67.05%	63.93%	46.86%
Date	20Nov02	14/3/03	15/4/03	9/5/03	end 1/7/03

Int, Upper-Int, Acad-Int # of Students

Student #	316	172	154	130	117
Date	20Nov02	14/3/03	15/4/03	9/5/03	end 1/7/03

Pre-Int, Int & Acad-Int Average roll size

Ss per class	10.19	10.75	9.06	10	9.75
Date	20Nov02	14/3/03	15/4/03	9/5/03	end 1/7/03

Pre-Int, Int & Acad-Int attendance

% attendance	68.10%	64.42%	64.03%	64%	52.82%
Date	20Nov02	14/3/03	15/4/03	9/5/03	end 1/7/03

IELTS Total # of Students

Student #	84	31	44	46	19
Date	20Nov02	14/3/03	15/4/03	9/5/03	end 1/7/03

IELTS Average roll size

Ss per class	10.5	7.75	8.8	9.2	9.5
Date	20Nov02	14/3/03	15/4/03	9/5/03	end 1/7/03

IELTS Average Attendance

% attendance	68.10%	89.03%	86.82%	78.7%	89.47%
Date	20Nov02	14/3/03	15/4/03	9/5/03	end 1/7/03

Academic Management and Academic Admin support allocation

All Classes, Total # of Students

Student #	527	394	362	305	241
Date	20Nov02	14/3/03	15/4/03	9/5/03	end 1/7/03

All Classes, Academic Mgmt Allocation

Equiv mgr #	4.71	6.5	5.28	4.07	3.91
Date	20Nov02	14/3/03	15/4/03	9/5/03	end 1/7/03

Academic Admin Support

Equiv admin #	5	3.7	3.38	3.86	2.31
Date	20Nov02	14/3/03	15/4/03	9/5/03	end 1/7/03

Responding to users

Prior to the usability sessions the intranet was set up on the computers of key staff. This was done on 30 August. A number of immediate issues came up, which related to the software and interface design. From the taped diary dated 30 August, 2002

> The intranet takes over a certain amount of space on the computer particularly if you reduce it to the smaller window view, which jumps to the top right hand corner of the screen. We controlled the positioning of the smaller view window ... I had to move a few of the users desk top file shortcuts ... In most cases that would be ok. One senior manager had a Liverpool logo set on his desk top, and he felt unhappy that the new intranet smaller view window partly covered his Liverpool logo. I negotiated to reset his screen resolution. ... One issue came up because one of the users found out that he could take over other people's identities, because we used staff pictures for the news page. He played with this and upset one staff member. We would have to add an optional password to the add news section.

We wanted to make the intranet a feature on user's computer screens, which they could not minimise or shut down. Therefore we locked the ability for a user to close the program window. From the taped diary dated 3 September, 2002

> We set up a competition with the marketing manager to see if he could crash the program by doing various things, so that helped us de-bug any stability questions. This competition drew interest from the owner ... who asked if he could access it from home. ... I sent him an e-mail with the address and password. ... We had to set up a first time password for the web site to unlock it. Once it has been typed a cookie (software memory file) is stored on the user's computer, so they won't be asked a second time.

Technical emphasis

There were a number of technical issues which came up in relation to IT security settings. The issues and solutions were of a technical nature, resolved by actions taken by the intranet development team and the IT department. A point however to note is the need for cooperation between the IT and new media staff. Our focus was on communicative and knowledge management aspects of the organisation, whereas the IT departments focus is on setting up and maintaining a robust IT infrastructure and security protocols. The main boundary crossover, where we needed to cooperate, was related to how the internal network connected to the Intranet, usage and login issues. There was also cooperation where, for the purposes of knowledge management, my interventions were to lead to changes in folder logic and user access on the shared network drive. Questioning of the key staff was important. Any new project needed the support of key staff.

1.3 The second stage: Usability testing of prototype

Usability testing is the activity of pretesting software or technology and using feedback to refine the product.

One reason for usability testing session feedback was to get buy-in to the project by involving key staff in testing the design of the intranet. At the same time I wanted to see the users discover what did not make sense to them in terms of intuitive use, benefits of the design, and to detect any bugs. Bugs have a negative effect once software is introduced and can lead to people giving up, so it was important to discover these early on. I observed the group of staff testing the intranet and collated feedback.

One reason for usability testing session feedback was to get buy-in to the project by involving key staff in testing the design of the intranet.

I had referred to a usability testing report by Molich (2001) on "230 Tips and Tricks for Better Usability Testing". Rolf Molich from Dialog Design in Denmark is a professional Usability Tester.

The usability testing sessions took the form of a number of groups of staff trying out the prototype intranet. Three computers were available, so where there were more than three testers in the group, two testers would use one workstation. They were asked to perform certain tasks and to speak out loud as to how they were finding it, their impressions, and where they were finding problems. Part of conducting these sessions was keeping a written record of the users' comments. It included a wrap-up group discussion at the end of the usability testing session. The emphasis of this feedback was on the interface design and functional bug issues. Ideas for adapting the design were also forthcoming.

They were asked to perform certain tasks and to speak out loud as to how they were finding it, their impressions, and where they were finding problems.

Three groups of 4-6 staff members took part in the usability testing sessions:
Group 1: The first group was made up of the senior management team covering Business, Operations, Marketing, Human Resources, Student Services, and IT. This group represented the staff I most regularly worked with, and I ran this session.

Group 2: This group was made up of the academic and management team including those from the two off-site franchises. This group represented the staff that were most regularly accountable to me in my role as Principal of the organisation. This session was run largely by my PA. I introduced the session and reappeared near the end of the summing up stage.

Group 3: This group was made up of a mixture of operational and administrative staff who use computers for administrative purposes. My PA ran this session.

Group 1 had a second session after some iterative alterations to the software.

Tasks included in the the usability testing sessions were posting news items, registering on the forum, adding messages and using the calendar. Some users would dive off into other uses depending on their interest, and this was not discouraged.

It was emphasised at the beginning of the sessions that the staff were not being tested, but that they were testing the software and that I would appreciate it if they vocalised what they were thinking. Because I was testing the software, they were told that we would not coach them, rather we would only assist if they got completely lost with what they were trying to do. This facilitation approach made it easier to determine design and bug issues.

Warnings from the Philosophy of Technology (Cont...)

Sometimes organisations operate behind closed doors as far as management is concerned. This creates a kind of cultural concealment or distortion that opposes emancipatory creative and free thinking. This is not what organisations need if they are to survive the fast pace of change that is needed.

Heidegger argues that it is easy to have an overemphasised focus on what technology does and can do, thereby losing connection with what is important.

During the session, notes were taken according to the categories: Problems, positive findings, suggestions from test participants, functional bugs and usage scenarios.

Comments were collated onto a form. The written instructions for the person taking the sessions was, "Interact with the participants and take notes as necessary while the task is in progress. The interaction might be as limited as reminding the participant to think aloud, or as extensive as an ongoing interview. Provide help only when it is clear that the test participant is unable to solve the task alone."

Comments from the wrap-up group discussion at the end of the usability training session were recorded according to the following categories: The three most important changes they would like to see happen, the three best things, the small things that are the most bothersome, and ideas they have for improving problems.

After the usability testing session, I met with my personal assistant and new media designer. At this meeting we determined the most important usability problems to address. Following this, screen shots were taken to record the iteration stages and alterations were commenced.

In the summary notes of the usability testing sessions below, a number of comments have an asterisk (*) before them. These comments demonstrate a developing awareness of how the intranet may impact the participative and collaborative culture of the organisation.

> **Warnings from the Philosophy of Technology (Cont...)**
>
> I hoped to develop the intranet as an emancipation tool rather than one which would lock staff into rigid and unsuitable forms of interaction. It is important to me that the staff in the organisation are free thinking individuals who critique the environment in which they work. The question, then is how the intranet can be designed to serve effective participation.
>
> Another theorist, Feenburg (1996), points out that the danger of technology is not so much its threat to human survival as the incorporation of human beings themselves into technology. The tail should not wag the dog. It is important that when creating new ways to communicate we do not replace embodied world communications, which have greater relational ties.

Figure 1.3.1 Summary notes of the Usability testing sessions.

Take notes on the following:
bugs (Functional problems)
Post a news item: Why not make the refresh automatic so news item comes up straight after entering it? Do the news items expire? Can it say when it will expire? *Add an event to the calendar:* The error message when one box wasn't completed wasn't clear. Why not make the time AM or PM only rather than being so specific – do we really care about a specific time? Can't use apostrophe. *Register for the forum and add a topic for discussion:* Why not give an instruction to enter first name only? What is a signature? I can't delete except by backspacing. The smiley option gives an error message.
design flaws (Logical Problems)
Post a news item: Not sure what selecting a campus means? Where is the refresh button? Campus is spelt campuse! *Add an event to the calendar:* Why not have the category, for meeting, visit etc at the top? Why not make the order of the selections should reflect the order of information when viewing the calendar? *Register for the forum and add a topic for discussion:* Why do we have to locate New Zealand? Why not by campus? Is there a character limit on my message? Why doesn't it return to the list after I send my message? Could there be a general discussion area as well as by department?

Warnings from the Philosophy of Technology (Cont...)

Whereas Heidegger (1949) and Feenburg (1996) pose warnings about the danger of being taken over by technology, a third theorist Marcuse (1964) offers another emphasis – that emancipation has already been usurped by the advancement of technological society. In Marcuse's view the position of technology in society has already dehumanised us. In other words, while technology offers opportunities to satisfy people's needs, in reality it has already taken over. We shall see.

Another philosopher, Foucault (1988), is concerned, like Heidegger, with how people place themselves in relation to technology. He notices how people care for themselves through communicative practices that have a meditative component - rhetoric, study, diary reading and writing, in correspondence and in dream interpretation. Foucault asks what the "right way" of individual verbalisation is.

I wanted to be careful to enhance a feeling of belonging for staff.

Where did people get stuck (disorientated)
Post a news item: One user kept trying to enter news item over the character limit? Can there be at least 4 lines?   *Add an event to the calendar:*   Doesn't tell you how to get out of calendar or how to view it!   An event can occur over many days – not just one!   What does the calendar do? Does it pre-warn?   *Register for the forum and add a topic for discussion:*   How do I make a new topic?   How do I write a message?
Suggestions for improvement
Post a news item: Why not make refresh automatic?   * Oh, I can select another person's photo! You can write want you want and say it's someone else saying it.   * How can you prevent sexist comments or defamation?   *Add an event to the calendar:*   Either include apostrophise or tell people they can't use them.   Bypass unnecessary info.   How can it be edited? What if Warwick is away!   It should say who put it on and when in small type.   Give persons name on calendar rather than event.   *Register for the forum and add a topic for discussion:*   I want the delete key to work.   Would like to be able to adjust the font size.   Using first name only is becoming an issue as the staff numbers increase!   *Other:*   The Usability session should not be live – it should be a controlled environment.
Positive comments
Post a news item: Looks like a chat room.   * Good for communicating globally.   Good as an additional tool.

Socio-political theories of technology

While these philosophical perspectives provided some "attitudes" for me as the project manager, there are also some other socio-political theories that I wanted to take into account for the design and implementation process. Writers such as Negroponte (1995), Winstone (1998), Anderson (1995), McChesney (1995) and Habermas (1991) focus on the wider socio-political ramifications of digital and Internet technologies.

Their socio-political theories led me to be very concerned to keep a user-centred design.

Winston (1998) presents the case that human decision making determines change. By this I took it that I needed to accept the intranet development process would be an evolving project and that the outcome could be quite different from the original intention.

Add an event to the calendar: No entries *Register for the forum and add a topic for discussion:* No entries
At the end of a session have a summary discussion to ask:
What are the most important changes you would like to see happen?
Post a news item: I would like to be able to update my photo. * It could be overused and become overwhelming. *Add an event to the calendar:* Would like to see the whole year, not just by month. Shouldn't have to choose by category. Should lead users through more easily. * Could see anonymous messages could lead to negative comments. *Register for the forum and add a topic for discussion:* Would like a font size option. Would like to be able to use the delete button. Would like the selections to look like MS Word as much as possible.
What impressed you the most?
Post a news item: Relatively user friendly, therefore no instructions needed. When would people check it? *Add an event to the calendar:* * This will work well with the franchises. *Register for the forum and add a topic for discussion:* Ease of use. * The concept of all users communicating across the organisation.
Think about the way you work and the way you communicate through the organisation. How do you think the Intranet serves the organisation's work goals and communicative culture?
Post a news item: * Bad if it replaces face-to-face communication. * Good if you can't find someone. Suggest auto-checking censorship/auto-cut for swearwords.

Social necessities and key benefits

Winston examined technological change from the telegraph to the Internet and points out the characteristics of change and the sociological factors that underpin them. He noticed that, no matter what innovations are possible, in the end it is the social benefits that ultimately shape the destiny of how technology will evolve. According to Winstone's line of argument, three things can evolve from a technological project – new technology, redundancy, and unexpected spin-off technology.

For the project, there would need to be key benefits for staff to spend their time in a virtual space as opposed to real world experience. For example, there would be no benefit in having dialogue on line if the dialogue can happen in the room next door.

Add an event to the calendar:
* If everyone had a computer, it might be relevant.
* If info about events pops up it could be good, but anyone can write about anything.
Register for the forum and add a topic for discussion:
* Potentially valuable if people see the discussions and respond. It can be frustrating trying to get people to sit down and talk about issues.
* More people across the organisation will potentially be able to understand the issues and may contribute their ideas, where otherwise they not normally do so.
* As the organisation expands this will help people to feel like it's a small company.
What other feature/ do you think might better serve the organisation's work goals and communicative culture?
Post a news item:
* Bad if it replaces face-to-face communication.
Good if you can't find someone.
Suggest auto-checking censorship/auto-cut for swearwords.
Add an event to the calendar:
Prioritising resources for teachers is more important than this.
Register for the forum and add a topic for discussion:
What if you want confidentiality.
Give instructions on keeping comments small.

1.4 The third stage: Prototype testing with key users

The prototype testing involved setting up the intranet on a number of key user's computers in order to improve it and get support from key staff. They would be using it in their normal work day and we would try and be responsive to their issues and adapt it as necessary.

These people were selected primarily because they were considered to be in close relation to me and therefore were amenable to support the project. These people included the owner of the college, The marketing manager, the operations manager, the student welfare manager, the human resource manager, the academic manager of one of the franchises, and an academic manager of one of the larger departments. These staff were chosen because I was most directly involved with them in my day to day work, and because of our strong ties, I felt they would be most supportive and thereby be patient with any debugging that was necessary. Once the intranet was set up on their computers they began to provide periodic feedback.

I would randomly show them how to use different features. I did this rather than setup training sessions for a number of reasons. Firstly, they had already attended the usability sessions, secondly I hoped that the interface design was intuitive enough that it could be worked out without assistance and they would discover how it worked and the benefits.

Social necessities and key benefits (Cont...)

Boyce (1997, p. 65) likewise comments that, "The social determinant for the up-take of technology is according to the benefits to the user". Boyce (1997, p. 63) compares this to the mainstream up-take of washing machine technology and the microwave. The release of time is the real benefit. The aim, then, is to determine the needs the technology might satisfy and to build the technology to meet those needs. Therefore, in designing an intranet for an organisation, questions about how staff already work will be significant. I had to expect staff to 'vote with their feet' in terms of what aspects of an intranet they might use or even, whether they would use it at all.

The idea of linking pictures of staff to any news items posted was successful. Staff like the idea that their photo is displayed with their news items on everyone's computer. From the taped diary dated 18 September, 2002

> I went out to one of the franchises, which is just about to open, and put the intranet on the Manager's computer and also the Director of Studies and showed both of them, how to add a news item in. This seems to be a very successful way of introducing the intranet, because people simply like to see their photo, and with their own news item. ... they get quite excited, that if they put a news item there with their photo, it actually comes up on other peoples computers. Technically it doesn't seem surprising, but people are surprised by it. They see a new way of communicating... it's a really good way to introduce people into this concept to start getting them on board.
>
> One of the franchises had made complaints about a lack of communication and support. ... it will be very interesting to see whether the intranet will be a mechanism of support.

Social necessities and key benefits (Cont...)

Turkle (1998) has written extensively about psychoanalysis and culture and about the psychology of people's relationship with technology, especially computer technology. She sees a world where the we will work with the embodied and virtual technologically mediated worlds simultaneously:

> We are not being so turned to the virtual and we are not going to the completely physical. We are going to learn to be between the two worlds. Already we see that when people meet virtually, they want to meet in the physical realm. Communities that begin in cyberspace start to grow in other places as well. And people who are off-line friends do more and more of their socializing on-line. It goes in both directions. (Turkle, 1998, p. 310)

Communication management

I was responsible for communications with and support of the franchises. There were a number of management strategies related to that, such as the franchise key staff reporting to a management meeting once a month at head office, setting up lines of communication for staff support across the campuses, setting up a professional development regime. My role did not involve business communication, which was handled at owner level an executive assistant.

I viewed the intranet as one mechanism for undertaking my management responsibility, specifically because of its potential as a communication channel and for disseminating manual information and forms, and for recording reporting data. Since the management activity for establishing the franchises was undertaken at the same time as the intranet was being developed and implemented, questions of how it supported communication and possibly file sharing and on line reporting was in the forefront of my mind. The degree that the intranet would factor into the strategic equation was being explored.

Social necessities and key benefits (Cont...)

The intranet, therefore, should offer a beneficial coexistence. McChesney (1995) likewise considers the social consequences in relation to parallel real world-virtual issues, asking:

> Is it really so wonderful and necessary to be attached to a Communications network at all times? ... Cannot the ability of people to create their own 'community' in cyberspace have the effect of terminating a community in the general sense? (McChesney, 1995, p. 145)

In designing and implementing an Intranet for an organisation, questions about negative social consequences should also be considered. Why or why not staff have taken or not taken to certain features of the intranet design should be explored. The learning about the organisation's culture throughout the project's evolutionary development may be as significant a spin off as any technological gain.

Feelings of belonging

One question has been whether the intranet would influence the culture of the organisation and one aspect of that is the feeling of belonging that it may enhance. From the taped diary dated 26 September, 2002

> *A key person was in discussion with me about the politics of the organization. The discussion covered the communicative politics and belonging issues that have to do with the franchises that we have just started. Comments were made ... about the success of the management interventions during the start up of the franchises. There was a perceived feeling of belonging and pride amongst the staff. The same comment was made about the intranet, how it gave people who were using it a feeling of togetherness.*

In terms of the project, the indication that the intranet, along with management intervention, was building a feeling of belonging is significant. From my perspective I recognise that staff need to communicate with each other. Mechanisms need to be created to encourage communication, so the feeling of belonging can be enhanced. The fast pace of change in the organisation brings with it stress and that stress is exacerbated if people are not working positively together. If a feeling of belonging is nurtured, then staff will work better together.

1.5 Taped diary of periodic feedback

This recorded data was collected through periodic feedback by users. I also recorded any relevant response either for adapting or debugging the intranet or of management interventions to support the project. The data reflects how the intranet was being shaped according to the organisational structure and culture.

Using this unstructured approach meant I was not imposing a theory or presupposition about what to expect. There were no formulated questions. The participants would provide feedback on the basis of trying to help me get the intranet working in a useful way in the organisation.

Once the intranet was debugged and some user interface alterations were complete, it was launched within the organisation by installing it on every computer. A number of initial set-up issues were solved and several training sessions undertaken.

The intranet was not launched for all the staff at the franchises, but only installed on the academic and centre managers' computers. This was because the franchises were semi-independent and it was therefore appropriate that these key managers try it out without any requirement on the franchise owners. At this stage, the product was still considered an evolving project rather than an

Social necessities and key benefits (Cont...)

Along these lines, Boyce (1997, p. 59) talks of the "easy entry into others' spaces". Just as Internets can make social contact in the world more accessible, it might be the case that social connections between staff are facilitated through the intranet. Social connection is relevant to the project, because there are four departments in the organisation and at the inception of the project, these departments were beginning to develop as separate social entities. The normal departmental social circles and the way the organisation constructs social gathering and staff meetings could impact on how an intranet affects the social connections of staff within the organisation. I was also going to manage how the meetings and reporting structures worked in the organisation. This is perhaps the uniqueness of this project. I wasn't just introducing and intranet to the organisation. I wanted to partner with the staff in evolving one that would help the us be more connected and able to participate more and I was going to try and change the other structures of the organisation to enhance this. I did by the end of the project have responsibility for a role I came to understand as internal communication management.

item of proven benefit for the franchises.

During and following these stages there was sporadic feedback of a varied kind. In general, staff were very helpful and I appreciated their input. Sometimes this would occur as staff approached me about problems or ideas for improvement. Sometimes comments would be made in the course of day to day business. When this happened, I might respond by promising to take certain action or that I would look into it, and this sometimes led to impromptu meetings. I kept a record of these meetings and my comments about the feedback. I kept a tape recorder in my office and would record these events as data for the thesis. For each recording, I would state the date, explain the event and any responses or planned action.

1.6 The fourth stage: Launching and training for all users

Rolling out the Intranet

The prototype went through a number of iterations and the software was debugged based on feedback from the usability session, as well as comments from the five stakeholders who had the prototype installed. The intranet was then ready to roll out to all the computers in the organisation.

All management and administrative staff in the organisation have a computer and teachers have shared access to a computer.

Once it had been in use for enough time for staff to form opinions a questionnaire was undertaken to examine opinions on it and how staff generally used their computers, how they accessed files on the network and how they communicated in the organisation (both through technology and face-to-face).

Potential of network digital technologies

Negroponte (1995) holds the view that digital life flattens organisations, globalises society, decentralises control, and helps to harmonise people. This is a positive approach. It was my hope the intranet in itself would be a tool for positive change - just as the process of evolving it and implementing it would be.

Likewise, Anderson (1995) similarly suggests that the technology itself creates cultural change. Users have potential unlimited and unfettered access to the Internet not because of any community, business or government will, but because the characteristics of the technology avoid measurement and control. According to Anderson, instead of people making a culture, the Net itself is responsible for defining a new creatively inspired networked communication culture.

Technical emphasis

We set up all computers so that on start up the intranet would automatically start and take up space over their desktop. This was accomplished by a script written by the IT personnel. However, we needed to see that each user, after logging on, completed the 'intranet logon'. When the tick 'Remember me' checkbox was ticked a cookie was put on the computer so that the next time the user logged on it would go directly to the intranet main page. This login requirement was needed because the intranet was residing on the Internet, and it provides a firewall preventing unauthenticated access.

The login system was however, an issue in that when staff logged on to another computer besides their usual one, they needed to log in again. This is because of the way Windows works, by storing user profiles on each physical computer. Teachers often shared computers so it had the effect that teachers were not using the intranet.

Figure 1.6.1 Login screen.

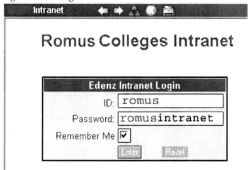

Once staff logged in they were presented with the intranet interface. This was achieved in May 2003.

Figure 1.6.2 Screenshot May 2003.

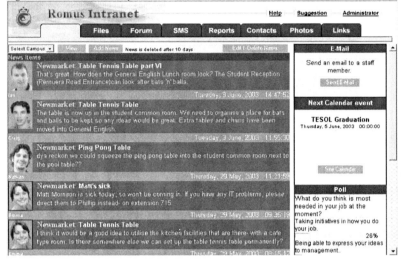

There was the typical Window "Minimise button' in the top right of the Intranet. If the users clicked it with their mouse it would minimise to the screen below and visa-versa.

Figure 1.6.3 Screenshot of the minimised screen.

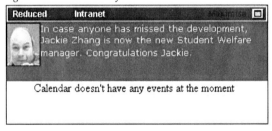

This minimised screen drew some complaint over a period of time and so it was culled so that the main intranet window minimised in the normal Windows way.

Communications, management and culture

From the organisational communication perspective, introducing the intranet would entail integrating existing forms of communications and making people aware of how the intranet can be one of the organisation's communication channels. From the taped diary dated 24 September, 2002

> *The newsletter that exists in an organization is just one way that news is dispersed, just on checking the latest newsletter, there were some items in it, like for example the Christmas breaks, so I picked up on that and sent an e-mail to the marketing department saying could you put those dates into the calendar. This is an example of encouraging people who are communicating with one system to begin to appreciate the intranet as another communication channel.*

Another point from the communicative perspective is how the interactivity of the intranet medium may reflect the communications culture of the organisation. From the taped diary dated 24 September, 2002

> *There is a fair bit of banter on the news page ... One staff members posted an item saying "Hey come on guys stop using this for banter." Now that didn't have to come from me, it was a self regulating thing, very interesting to see the office communication, sort of that friendly culture of Romus a bit of banter going onto the intranet as well, and even a bit of a self censoring from the staff - wouldn't want to stop that, just watch it.*

At this point the News page on the intranet is establishing its own communicative culture. There is a mixture of official news flashes and social information. This reflects the friendly atmosphere which exists in the company. The following screen shot reflects this trend.

Figure 1.6.4 Screenshot of the intranet September 2002.

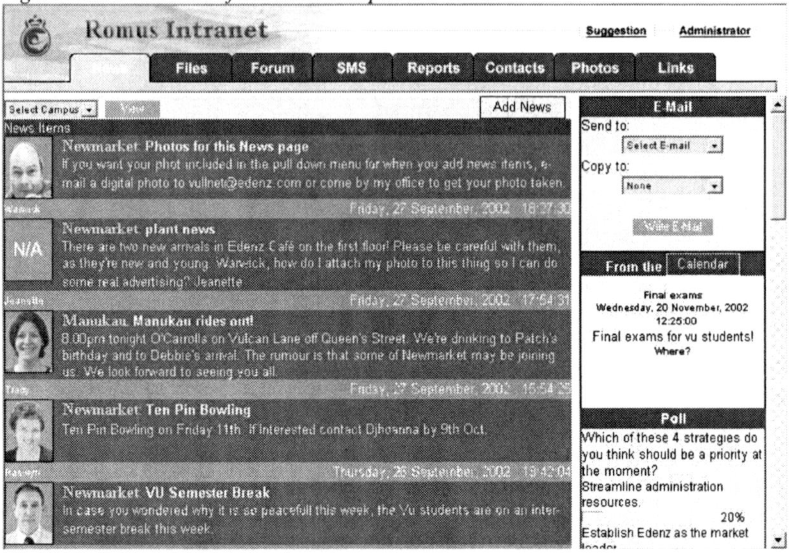

Another management intervention was to ensure there were enough computers for teacher access and to post instructions above the teachers computers on how to log in, post news, and use the forum. The intranet and this setup and instructional information was also presented at a monthly teachers meeting.

Training sessions and intranet uptake

The first training that was undertaken was to visit staff and show them how to enter an item on the news page (see News tab). There were no access restrictions and the user's photo would come up with their message. This training was on a one-to-one basis and it was done casually as I encountered staff in the organisation. From this point, staff began to use the news page.

As explained above, the franchise schools were rather tenuous in their association and as a first strategy we installed the intranet on the top managers' computers and on the academic manager's computers. I was in contact with these people directly and therefore had a working relationship with them for taking this initiative. The top managers had been involved with my franchise implementation team and been trained at the College and the academic manager's had been employed at the Colleges head office campus prior to taking up their positions with the franchises. Because of the internal politics associated with these franchises, I doubted whether the intranet would catch on, beyond these key staff. The ramifications of the politics, would likely impinge on staff loyalty and communicative connection.

For the forum (see Form tab) I co-opted a student, Alice, who was doing part-time work to run a training session. The owner had asked me to involve Alice in some kind of project. My idea was to supervise the process. She organised a meeting for a group made up of operational staff, registration and administrative staff. There were approximately six trainees. They brain stormed ideas for improving their work systems and we were to train them in registering on the intranet forum. The plan was that they then add their comments on the forum and I would see that some action was pursued and feedback provided via the forum. Unfortunately we encountered bugs in the software at this point and had to postpone this stage of the training. Unfortunately, Alice went back to full time study before this could be revisited. We planned to fix the bugs and reconstitute the group for another attempt. In the mean time other business priorities pulled my attention away from this endeavour. Later on, there was another attempt with the involvement of my personal assistant the quality assurance manager and with a new forum. This proved, however, to lack impetus. The entrepreneurial culture of the organisation listened to staff that were assertive in showing initiative. These staff could make themselves heard without the need for an intranet. Other staff, who were less assertive, saw little reason to contribute ideas that they felt would not be seriously listened to. This issue was to spur me on to take management interventions to redress the concern that we were not including staff as we should. For example, we began to organise monthly full staff meeting, and rotate opportunities for staff, not normally given the opportunity, to talk about what their department or area was up to. I came to see this intervention as falling under the role of communication management.

Technical emphasis

The Colleges' Student Management System (see SMS tab) enables student registration details to be stored. I simply met with staff that used the SMS in their normal course of work and showed them how they could access the SMS through the intranet. The SMS works via the internet in a similar way to the intranet. Data is stored with an external Internet service provider, using a data base on their server. The Colleges pays a licence fee to an external contractor to manage the software and its upgrades. It was clear at that point that the intranet was providing no real advantage for users who accessed the SMS. They were used to accessing it with their Internet Explorer browser and continued to do so.

The reporting that was envisaged for the intranet (see Reports tab) was initially set up for two areas of the college. One area was to keep a record of student welfare data and the other was to record the number of students in the ESOL department and the management and administrative staff hours allocated to run the programme. The first data was entered on November 2003. The idea was to incorporate accountability feedback according to historically proven reporting requirements.

For the ESOL management team, I undertook a training session to explain the principles behind the intranet system. The goals were based on agreed teacher, management and administrative staffing allocations according to student numbers, which reflected fair resourcing and provided for budgetary accountability. It was explained to the ESOL management team that they were able to make their own staffing decisions based on the goals, as long as they would complete the report on a monthly basis and that the programme leader would explain the report to the owner once the data had been entered. At the same time I undertook a training session with the owner on how to access the Report area of the intranet.

Figure 1.6.5 Screenshot of the Academic Allocation report by January 2003.

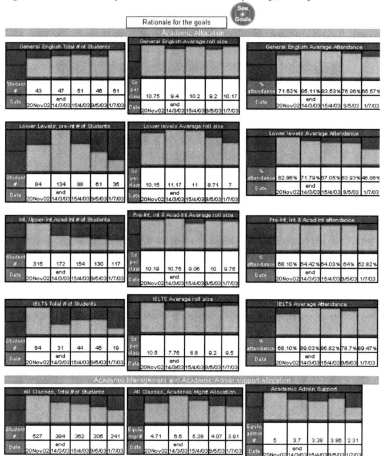

After training the ESOL management team about the system and how to enter their data, there was also the question of what parallel management activity should happen to see the system incorporated operationally. From the taped diary dated 22 April, 2003

> About the report procedure for the intranet - just following up on the process. The data has been entered. Just to double check with the administrative person who collated the data ... to check that...the guy responsible for the academic area for which the report represents ... to give a report to Thomas the owner, thus trying to move this process into the realm of normal culture – a system of monthly accountability. I am nurturing that through and there is a data inadequacy where the student data is taken off the SMS and that is 2 weeks old, so I am going to have to check on that side of the system so the data is true.

This however, represented only one of the departments of the College. Efforts to create the same system for other departments were not realised. This could have been because the ESOL department was much larger than the others and had an administrative person that could be allocated the task to enter the data. Later, significant integration of sub-departments and layoffs meant that the system as described fell into misuse. When an organisation has a high rate of

Potential of network digital technologies (Cont...)

With intranets it is not as clear cut. I have considered that the technology is not exactly the same for the Internet and intranets. On the one hand the technology is similar, but on the other hand an intranet falls within an organisation's control through the role of communication management. Any communication, since it would be open would be censored or self censored by the same terms as any existing embodied world management censorship of the communicative culture.

However, there may not be a true forum for open critique. When this idea is applied to organizations in general it can be asked whether decision making is undertaken behind 'closed doors' and whether provision for client and staff forums, including those mediated through an intranet, provide valued critique which management responds to or whether management largely ignore it. It can be expected that the stance of leadership on encouraging or excluding participation will affect attempts to apply a user-centered approach to intranet design. We shall see.

change, it impacts on changes in the way the reporting systems work! We could not keep up with the pace of change necessary to manage reporting through an intranet engaged system.

> *When an organisation has a high rate of change, it impacts on changes in the way the reporting systems work!*

The contacts page on the intranet (see Contacts tab) was designed to replace an existing spreadsheet which contained staff details. Staff could edit their own details with managers being able to add new staff and change the position and 'Contact about' details. This contacts page proved to be very effective. One reason was that as peoples roles changed that could easily be updated.

Figure 1.6.6 Screenshot of the Contacts page.

One aspect of the development of this contacts page was its relevance for the franchise schools, and the ways in which they wanted to be connected with staff at the main campus, who were working in similar fields. From the taped diary dated 28 November, 2002

> On going out to one of the franchises and discussing the communication and support issues it became evident that they don't want so much one person working as an intermediary. They preferred to be put in touch with their colleagues at the other schools. At that time the contacts page did not include enough information, so consequently we decided to adapt the page to add a section of what to contact people for. We also added an edit button to make it possible for staff to change certain details themselves.

The photos page on the intranet (see Photos tab) was designed to provide a communicative belonging. The idea was to put photos into sub-folder sections and a menu system would self-generate. I hoped that staff would find their way there. Also if it was to succeed we would need to take up to date pictures and upload them for the intranet on a regular basis. I was without assistance for some time and was overrun with the pressures of a boom period of operation and stress from internal politics. The job priorities meant that my focus was averted from maintaining operational updating of photos or of training staff in this area. A part time worker in the marketing department had taken numerous pictures around the organisation and stored them for access, but he had devised his own system for this. After they left the organisation I located the pictures on the network drive and incorporated them on the photo page of the intranet. This situation where a staff member undertook a task and devised their own system without any real collaboration was endemic in the organisation. The problem with this section is someone needs to be allocated the responsibility to keep it updated. In our situation this was difficult.

The links page on the intranet (see Links tab) was designed to provide an alternative to common links where staff normally used their Internet Explorer browsers. It also was a nice way to integrate the website of the Film Department.

Communication management

Once the intranet had been introduced a number of initiatives were required which came under the general role of communication management. Up until this time there had not been a specified role in the company.

One activity of communication management was to keep channels of communication open and this included any mechanisms fir those channels. From the taped diary dated 23 January, 2003

> The intranet new media developer pointed out that not many people had been using the intranet in the New Year. For example, people were using their diaries rather than the calendar. We asked ourselves how we could create significant advantage for staff to refer to the intranet calendar. Perhaps we could get it to automatically send out reminders - useful perhaps. We realised that unless we did that and offer it as a feature it was going to be hard to convince people to use the calendar.
>
> One of the franchises commented that there were no news items. We realised that the New Year was going to be an opportunity to form habits and so we put some news up as a prompt.

Another activity of communication management that was becoming clear was that of recruiting key staff to assist with communication networking. From the taped diary dated 18 February, 2003 a simple comment reiterates this theme.

> Trudy was asking how to use the forum. We need to try to get back to people like this because they are an asset.

The staff that were providing feedback and supporting the intranet were important because they had bought into the idea. They appreciated that communication is being strategically worthwhile and should be encouraged.

Another activity for communication management was ensuring that staff had access to the intranet as one channel of communication. The first part to this was to ensure staff could access a computer and the second was to ensure they could access the intranet on it. If this was not achieved then some staff could be locked out of the intranet communications channel. When teachers did not have primary access to computers, computers were set up in the teachers' room so they could easily access the intranet. From the taped diary dated 18 February, 2003

> One of the supporters pointed out that the intranet had not yet been installed on the computers for the Business and Computing department.

Similarly we needed to look at how the intranet started up with its log on screen and overcome any barriers where staff were getting stuck at this point. From the taped diary dated 17 March, 2003

> Jason is feeding us information on how people in the Business and Computing department were responding to the intranet. When people logged on to their computers they were confronted with the login screen for the intranet. Because in many cased they had lost the instructions they were simply minimizing the logon window. We needed to resolve this. The reason we have the log on in the first place is because it protects the organisation from hackers. ... We went through the organization and unlocked any that needed a password. However, sometimes there were two users logging in to one computer and so we did another sweep the following day.

Another activity of the communication management role is to ensure that information is up to date. From the taped diary dated 17 March, 2003

> Jason pointed out that news deleted every 30 days makes the news too old. They wanted the time reduced. Staff were getting sick of seeing old news and likewise staff that put news on don't like to feel people were getting bored seeing their news item. ... We changed it to 10 days before news was auto-deleted.

Another activity of the communication management role is training. Where there is any technical medium, training is required. This is exacerbated where user participation is required. From the taped diary dated 7 April, 2003

> We are working on news and contacts page issues. We are trying to make a print version of the contacts page better than the old system. We are making it so that the photos also print so if it is stuck on the side of a computer you can see the photo of everybody. We are also adding features so users can edit and delete themselves. In the case of news we will let everyone edit and delete. In the case of the contacts page we will let managers do it. We will need to train the managers how to do it, except for photos – our function will be to do that because it is too technical to expect them to do that.

There were numerous other training points, because we were trying to get staff to generate content rather than create it ourselves. The training was beginning to pay off. The first breakthrough was when we noticed that the idea of staff adding their own news caught on. From the taped diary dated 9 April, 2003

> Double people are putting news items on than had been trained!

From the taped diary dated 26 March, 2003

> The next training meeting on report forms is scheduled, so I can show the ESOL management team how they can complete the report and how their responsibility relates to it.

From the taped diary dated 31 March, 2003

> There was a training session with owner on the intranet, going through how to put news on again, how to access the ESOL

> *report feedback as part of the intranet and about the calendar part. He had some points on bugs he had noticed and so we are working on those straight away.*

The extent that staff were using it was beginning to be incorporated into the communication culture, as can be seen from the taped diary dated 10 April, 2003

> *Won another person over - another person has made a comment that "I used to use it - took it for granted – now I think it is good".*

Another activity of communication management is to see that staff feel a part of the culture – that they feel included in the communication. People do not have to be a part of every bit of communication. They need to feel connected in a way that they feel they have something to contribute or need to know. From the taped diary dated 10 April, 2003

> *The discussion we are having is about the difference between managements use and teacher use of the intranet. Do teachers feel they are part of the organization? ... We think they would be more interested in the forum than in adding news items. We thought to ... add a button so teachers can send email even if they do not have an email account. At the moment staff have email but not teachers. By adding this features all the features are available to teachers.*
>
> *We are also making it so anyone can add themselves to the contacts page. They may not want to but it means they can include themselves if they wish. We did this because ... although ... the admin staff do not need the teachers on a phone list, if the teacher wants to be on the list - lets make an open system where anyone can be included.*

We hoped that key staff we were recruiting to assist with communication networking could help us to get the forum going, thereby connecting people up. Another role of communication management that was becoming clear was that of recruiting key people to assist in ways in which we could create staff discussion for the improvement of the organisations services. Setting up staff training days might be a way of handling this. I recruited a part time staff member to look into this as a project. From the taped diary dated 1 April, 2003

> *Alice who works part time was asked to work with me on taking initiatives in creating communication activities. We discussed what strategies and tasks we could come up with to improve customer/client contact for successful encounters. We wanted bottom up staff involvement. The approach was for Alice to go to the key managers of departments and ask for time to run brainstorming session with staff groups. ... In terms of the intranet we would create discussion based on issues which are faced by them at the point of contact with customer and clients. We would then train staff and carry these discussions on to the forum.*

The forum though did not show much promise. We hadn't yet got over the hurdle of training people how to register and there were a few bugs because of how the company firewall was working in relation to the forum. Governing the work pressures there seemed too much effort required and there was no pressing need in the organisation for this kind of discussion forum. This lack of inertia shows, from the taped diary dated 15 June, 2003

> *Bruce had been shown how to use the forum, but it wasn't coming off, because the person I was using for training in this area disappeared out of the company. This person had been only part-time. She had now got tied up in her studies and had not been seen for some time. Bruce said, "Hey, what's happening. Some people had looked at the forum and there was nothing happening - there were no messages". Bruce has just taken over the library so I have just made a new forum section called library. Bruce is again one of these key people who are supporting the trials of the Intranet. These people are behind you. ... and it is really good to support these people.*

Another idea to try and get something out of the forum was to make a section for staff to make comments about the Student Management System (SMS). Since a 3rd party developer was responsible for it there tended to be ongoing criticism of the needs we had for the software to suit our purposes and the developer to willingly make these changes. From the taped diary dated 16 July, 2003

> *I met with Jonathan the 3rd party developer of the SMS system. I mentioned the idea that I could get the users to put their ideas*

> on the forum of the intranet and that he could also access it because the forum is online. What I am going to do is email him and ask him to introduce himself on the SMS part of the forum. Then go to the people in the company who do use the SMS and ask them to add suggestions for changes.

Knowledge management

For the Files location (see Files tab) a self generating menu creates links to the files in two folders on the common drive – Common and Manuals. The manual folder is where policies, procedures and control documents are located. The Common folder is where commonly referred to documents, such as job description and pay policies – human resource material, is located.

Figure 1.6.7 Screenshot of the Files page.

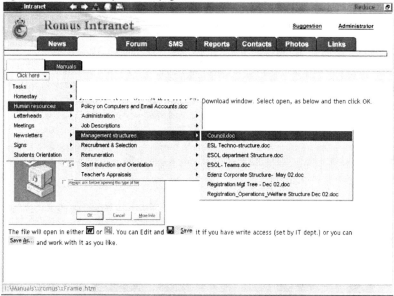

The idea is to make these documents more accessible to staff in the organisation. In order to do this I undertook discussions with all departments and established the filing conventions of the two main folders called Common and Manuals.

Historically the organisation had manuals developed by an outside consultant and these were formatted along the lines of ISO 2000 conventions. The ISO 2000 model was based on quality control documentation for a production environment. This meant that historically the organisation's documentation was overly cumbersome and had fallen out of use apart from one manual which the President used. While these manuals contained many useful policies, procedures and control documents, they did not represent the way the company was working in 2003. As Principal I was responsible for re-structuring these manuals in line with our current practices.

I allocated the task to one of our administration staff to go through the documentations with each department head. This eliminated many duplicate versions and determined which sections were pertinent. The manuals existed in hard copy and one intervention was to break these up into named sections so that they were more accessible via the computer interface. This resorting of our manuals was undertaken at the same time as the development of the self generating intranet menu. Staff were being reminded of the presence of these documents.

However, we were still far from creating a system whereby the manuals would become active documents for the organisation. One problem was that departments worked mainly in their department folders and so did not naturally refer to the manuals folder on the shared drive. It seemed more appropriate to break up the manuals from a centralised location and put them into each department's folder. Arriving at this conclusion meant the idea of making the files more accessible through the intranet was becoming redundant. By putting the manual files into each department's folder it achieved it better than the intranet interface idea.

These endeavours as management interventions illustrate how the development of the intranet generated activity which as a 'spin-off' was significant for the knowledge management of the organisation. The efforts created a learning curve in the organisation on the placement of files according to an agreed system. Rather than training staff to access files via the intranet the endeavour led to better knowledge management within the organisation.

Since I was trying to construct the intranet along the lines of user-centeredness, there was a parallel strategy to try and get departments to create their own manuals. From the taped diary dated 7 & 11 May, 2003

> *I have developed a text on how to write manuals – sort of a hyper linking system using Word. And so I will take it to Julie, who I have heard has been reshuffling files. I then met with a couple of other key staff on developing manuals to try and get a consistent system.*

I then conducted a training session for managers and staff likely to be involved in manual writing. This was to try to encourage changes in how we develop manuals according to a consistent format. The other part of the training session is to familiarise key staff with Romus filing systems and conventions.

Knowledge management ideas

Other ideas were explored in terms of how we could record various activates. One example of this was student evaluations. These are when we ask students to evaluate the services provided, much like a customer services survey. I asked myself how we might manage this knowledge and whether the intranet could be used for this? While this idea has not been taken up the intranet is a mechanism that could be used for such things. In this case, there would need to be a project to redesign the student evaluation forms for each department and then to set up and monitor the system. Because of work pressures this has not been looked at further. However, it is a good example as to how the intranet can be developed to manage the knowledge gained within the organisation. From the taped diary dated 17 December, 2002

There was a discussion with the IT developer about whether we could handle student evaluations on line. We discussed whether there was a way of doing this so that it was significantly better than how we were presently handling it. Previous reports produced following by collation of student evaluations were time consuming. These reports included graphs as a way of summarising the findings. We discussed setting up an on line student evaluation form. Once the data was recorded on a data base, we could then use the Internet to automatically create graphs, thus reducing the time that was presently involved.

The purpose of the reports is to provide feedback to management so that they are aware and initiatives can be taken to improve the services and facilities.

At the moment the student evaluation forms are completed by students and the forms are kept on a desk. Maybe as a bundle it moves around a department, where staff are supposed to look for critical issues. Overall, there is no organised way of ensuring improvements are made to the organisation. It would be extra over what people normally have to do to survive. It is the sort of thing that gets overlooked regularly. ... we would still need to get student to fill the paper based form out and so it would still require someone to enter the data. ...if we have the mechanism we can look at different ways students can do it.

Relationship management

Another aspect is relationship management. The relationship between the franchises deteriorated over the first six months of 2003. Primarily this was a business issue where the franchise owners were seeking independence. Communications broke down completely in one case, and remained on an academic level only in the other. This naturally affected the intranet in so far as it extended to the three geographical locations, and meant that it became increasingly difficult to maintain mutually beneficial working relationships operationally. 'Power plays', whether for position attainment, status or for ownership are often experienced in the workplace, and can significantly affect relationships and communication. For those not directly involved in these power plays, however, there are always opportunities to strengthen relations and communications. Sometimes this will involve unofficial agreements between parties. From the taped diary dated 15 December, 2003

> ... about the communication between the franchise and us ... it is also about the relationship because I notice on the Intranet there is not really communication from the other franchises sites. ... If we felt connected and integrated we would see some messages coming from the franchise campuses. ... I had a conversation with Trudy and I talked about the conflict that the organization has because of John. Trudy is caught up in this conflict because they are trying to work with us on a managed academic level. But also John gives them certain instructions. Some times on academic matters without any consultation ... So there is sort of an issue. ... [the comment goes on to explain how the issue breaks down the communication] Better communication requires encouragement from both our sides and Trudy and I have now decided to do that, within the extent of our influence. ... [the comment explains some action we took to improve cooperation between our respective homestay and social activities staff]. ... [the comment goes on to explain how there are always a mix of staff on the spectrum of being willing to cooperate on a collegial level and that this would reflect on the Intranet].

1.7 Conclusion

I drove the design and iterations of the intranet, collate the data, and manage the parallel change initiatives in the organisation toward a more collaborative and participative culture.

The evolving nature of the development project meant that feedback initially informed the interface design and functionality, later it addressed integration issues, such as how users accessed computers and how different features were useful or otherwise in relation to how the organisation worked. In the final implementation stages, feedback increasingly addressed the issues of the culture, of the networked organisational model and collaborative and participative involvement.

The management interventions have been described, as they influenced the project. A number of spin-offs became apparent; one being that communication management has been an evolving role reaching across boundaries in the organisation. The reporting structure for the organisation was remodelled, as was the way the organisation's manuals were accessed.

One of the exciting events was when, after some training, staff started using the news page. It had caught on. In particular the way the users' photos were matched with their news items made it seem very personal.

During this multifaceted activity of the project, certain trends were emerging. One of these was the potential need for a communication management role in the organisation. Activity associated with this included the need to recruit staff to assist with communication networking (with supervision of user-centred activity in some cases); ensuring staff had access to the mechanisms of communication - computers and the intranet on those computers; ensuring that information was updated by users; training for technical mediums, especially where user-centred input was required; and ensuring staff felt part of the communication system as a whole.

> *certain trends were emerging. One of these was the potential need for a communication management role in the organisation.*

I was finding that it is the communicative aspects of the intranet which were surviving the rigours of change and restructuring. Along with this, the evolving parallel communication management role was proving to be the most relevant management discovery.

SECTION TWO

REVAMPING THE INTRANET

2.1 Introduction

This chapter focuses on what I was trying to achieve when revamping the intranet. The story is told sequentially, interspersed with interview feedback and includes explanation of how management interventions affected the process of implementing the intranet. Likewise, the story explains how the evolving intranet impacted on the organisation. This will complete the development story.

2.2 Revamped intranet: Interface and features

Dec 2003 - Dec 2004

After the intranet had been running for six months in the organisation, the staff had formed certain views of it. Staff had not been compelled to use the intranet and so the test had been whether, and to what degree, it was perceived as useful. At the same time, staff had discounted or ignored certain parts of it and they had developed opinions justifying their attitudes. I had received a range of feedback reflecting on the issues of communication. The feedback ihelped the revamp.

A number of screen shots of the revamped intranet are included in this chapter. Below is the logon screen with a simplified question, which everyone in the company knows. There had been problems where staff complained at not remembering the login password. Staff can guess the logon rather than have to remember yet another password. As long as they tick the 'Remember Me' field, they will not see this message again.

Figure 2.2.1 Login, which does not show again if the user ticks the 'Remember Me' box.

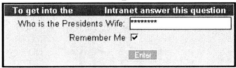

The main window is then presented on the user's desktop.
Figure 2.2.2 The main window.

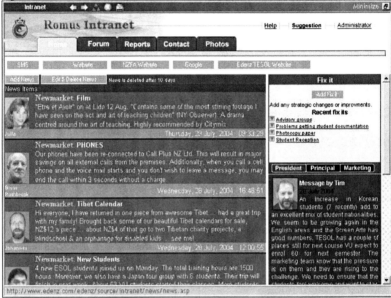

When staff click on the 'Add News' button, a window appears allowing them to enter the details, select their photo and submit it. It is then posted to the top of the list.

Figure 2.2.3 The 'Add News' form.

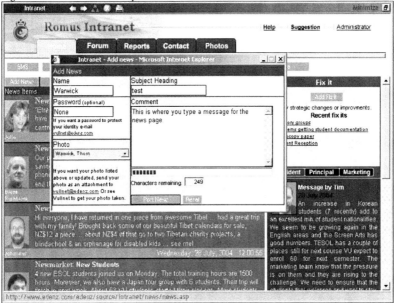

If a staff member clicks on the 'Add Fix it' button, a window appears allowing them to enter the details into the problem and solution fields, select their photo and submit it.

Figure 2.2.4 The 'Fix it' form.

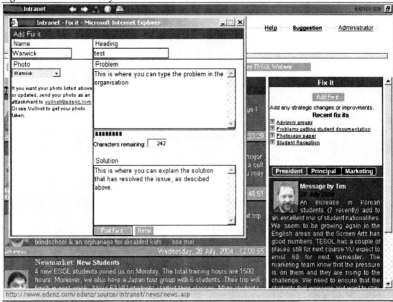

Once a 'Fix it' is submitted, it is posted to the top of the list in a similar fashion to the News items. The window below shows the most recent 'Fix it' entry. At this stage I was making all the entries. I had the intention to involve other staff in posting their own entries. The 'Fix it' system is a quality assurance initiative to record and publish how the school addresses strategic organisational problems.

Figure 2.2.5 The 'Fix it' page.

If the President, Principal or Marketing Manager click on their picture, this window allows them to update their message to staff.

Figure 2.2.6 The 'Add Message' form for the President, Principal and Marketing manager.

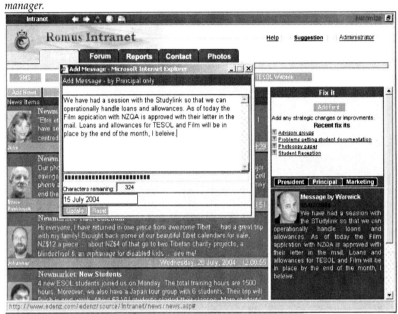

If managers click on the 'Contacts' tab and then the 'Add Contact' button, they can add a new staff member and the details to the list.

Figure 2.2.7 The 'Add Contact' form.

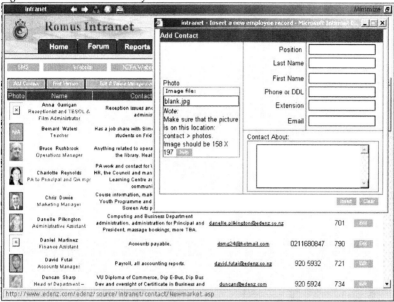

If staff members click on the 'Forum' tab it will come up, as seen below. They then need to additionally log into the forum. We standardized the login names to the ones staff normally use to log in to their computers. The password is the same for everyone, unless they nominate to change it. However, this feature was not at this time being used and it was later scrapped because the organisation was small enough that face-to-face discussion was considered far more worthwhile.

Figure 2.2.8 The 'Forum' page.

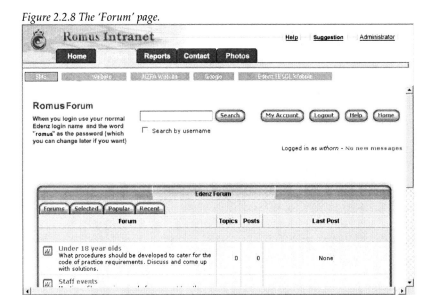

During the project, the organisation experienced significant growth and decline. The organisational culture had been distinctly divided into four departments, plus senior management and marketing staff were intermingling and networking more. Also, the franchise schools were now independent, so the need to unite staff across geographical boundaries no longer existed. The idea of using a forum for discussion was not seen as beneficial in this situation.

The benefit of a forum is that it can open up more diverse communications and it can keep a history of those communications. However, if staff feel they are able to communicate in other ways, to solve problems satisfactorily, then they will probably not use the forum. My strategy for the design process was not to be directive in its use, so when I saw little interest in the forum I accepted that the benefit was not significant enough to warrant further training. I think that minimal prodding and training should be necessary, because otherwise I would have to heavily coerce staff to use it. The principle of user-centeredness is that the intranet should significantly improve communications compared with other communication channels. Staff would therefore use the features on the intranet that they found most advantageous. They would do this without much direction, as long as the intranet was presented to them on their desktop.

I have previously discussed how a new emphasis on communication management evolved during the course of the project. One task I took on was to organise a monthly staff meeting, where we would organise for different staff to talk about their jobs. This meeting format opened up a better general feeling of community within the organisation. In fact the meeting focussed on company communications and that was because of the intranet project. Another way of seeing how the impetus of such a project can affect the culture of an organisation. Now I as a manager are starting to care about internal communications. This also explained why the need for the forum didn't really exist. We were more interested in communication face to face.

The same issue existed about the reports page – the intranet project had focussed us on how we report and we became interested in improving that which ended up being a non intranet event. By July 2004, the organisation's reporting culture had evolved into a monthly council meeting where management presents reports. The reports come form all sections of the organisation, are presented by senior managers to the rest of the management team, and represent the business, marketing, academic and strategic perspectives of the organisation. Most of the reporting is done in groups to enhance collaboration. This meeting consistently gets the nod from the President as the most cost effective and efficient reporting session the organisation has. It was far better than any attempt for intranet based reporting. As a result of the improved council meeting, the reporting feature on the intranet became redundant.

The feature under the 'files' tab has been removed. This feature enabled direct access to the organisation's manuals through an auto-updating menu system. This drop down menu and sub-menu system made it easier to find files, compared to the standard Windows interface. However the Windows interface also had the advantages of providing the means to manipulate folders and files and create short cut options. The intranet method was not significantly better. The re-ordering of the filing structure was undertaken to make the intranet menu better. The improvements existed for accessing files through the Windows interface just as much as through the intranet and had the effect that staff referred to the files more and were more aware of the filing protocols. This was a spin-off benefit, but it had no effect in enticing users to the intranet file menu system.

Figure 2.2.9 The update by 2005

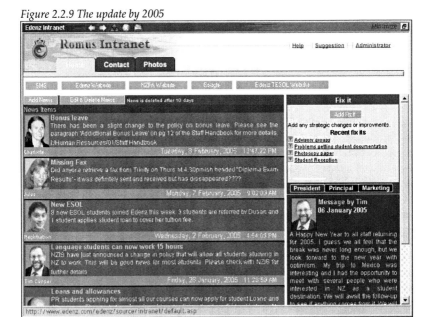

Figure 2.2.9 The update by 2005 (Cont...)

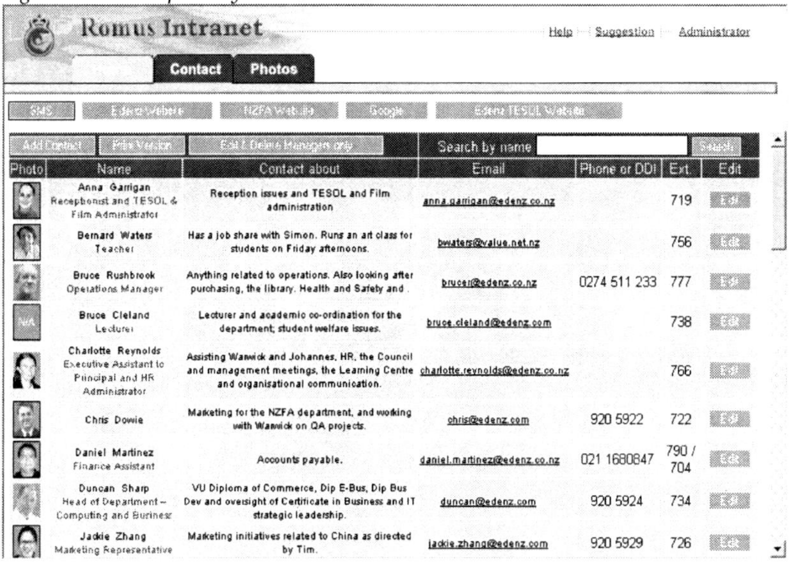

Photo	Name	Contact about	Email	Phone or DDI	Ext.	Edit
	Anna Garrigan Receptionist and TESOL & Film Administrator	Reception issues and TESOL and Film administration	anna.garrigan@edenz.co.nz		719	Edit
	Bernard Waters Teacher	Has a job share with Simon. Runs an art class for students on Friday afternoons.	bwaters@value.net.nz		756	Edit
	Bruce Rushbrook Operations Manager	Anything related to operations. Also looking after purchasing, the library. Health and Safety and .	brucer@edenz.co.nz	0274 511 233	777	Edit
	Bruce Cleland Lecturer	Lecturer and academic co-ordination for the department; student welfare issues.	bruce.cleland@edenz.com		738	Edit
	Charlotte Reynolds Executive Assistant to Principal and HR Administrator	Assisting Warwick and Johannes, HR, the Council and management meetings, the Learning Centre and organisational communication.	charlotte.reynolds@edenz.co.nz		766	Edit
	Chris Dowie	Marketing for the NZFA department, and working with Warwick on QA projects.	chris@edenz.com	920 5922	722	Edit
	Daniel Martinez Finance Assistant	Accounts payable.	daniel.martinez@edenz.co.nz	021 1680847	790 / 704	Edit
	Duncan Sharp Head of Department – Computing and Business	VU Diploma of Commerce, Dip E-Bus, Dip Bus Dev and oversight of Certificate in Business and IT strategic leadership.	duncan@edenz.com	920 5924	734	Edit
	Jackie Zhang Marketing Representative	Marketing initiatives related to China as directed by Tim.	jackie.zhang@edenz.com	920 5929	726	Edit

Figure 2.2.9 The update by 2005 (Cont...)

2.3 Revamped intranet: Views of staff

| Dec 03 – June 04 |

The interview data, provided feedback which influenced the revamp decisions. Data from the interviews gave me an opportunity to establish in my own mind where the value of the intranet lay in terms of serving the organisation. There were some very affirming comments. The negative slant to some comments reflected difficulties the staff were having with systemic issues and communication in the organisation, as well as the frustrations involved in trying to develop and implement an intranet.

Interview feedback

The questions in the interviews tried to uncover the issues . This interview data consolidated a view of the organisation as a modern organisation embroiled in a knowledge-based global economy, where there is a high rate of change. The interview questions also tried to explore whether the process of developing the intranet had a radical influence on the organisations culture, communication and knowledge management in a participatory way.

> *This interview data consolidated a view of the organisation as a modern organisation embroiled in a knowledge-based global economy, where there is a high rate of change.*

This data completes the story from the point of view of the users and those involved in the project. The interviews were conducted between October 2003 and June 2004. The first data set is then presented for the period 21 October, 2003 to April 2004. The second data set, from interviews conducted on 6 June, more directly enquires into how it has helped the organisation become more participative.

Collated comments from the interviews conducted between October 2003 and April 2004:

Q1 The intranet has attempted to do a certain thing...what is your understanding of those things and how successful has it been in achieving that?

> People are using it - adding news and contacts. They don't use files. Vaughn 11 Dec 03
>
> I understand about half of it, or am aware of about half. These have to do with enhancing communication, especially across departments. I.e. news and contacts. I believe it has been successful for those who also use/read them. Shirley 1 March 04
>
> Forum and contacts page not used. Contacts info is not as up to date and we use the phone list. Jane Liu 26 March 04
>
> Got it so we can communicate with people in the organisation directly – specific to administrators of the organisation. Doesn't include what might go on a webpage to persuade our clients. Intranet is source of disseminating info to organisation and encourages and is direct. Thomas 7 April 04

These comments reflect that the communicative aspects are the ones that are being seen as useful. It reiterates the rationale for cutting out sections as discussed in the revamp section above.

Q2 A number of staff use the news page of the Intranet. Do you think the mix and content of these messages says anything about the Romus culture? (Eg. Do we value social mingling of staff, do we believe the clients or administrative tasks are more important, do we assume that hearing about changes can come from different quarters?

> News, Contacts and Poll a little. Because it's in their face – they can see it – don't have to look. Photos make it personal. Vaughn 11 Dec 03
>
> Contacts – because it's easy to access and read. Shirley 1 March 04
>
> Notice Board – No other way for people to know what's going on eg franchises. Jane Liu 26 March 04

> Gives updated instantaneous report on staff emails and positions and telephone extensions. Another place is that we can access latest documentation. Quite good. Strength is that I can communicate with everyone instantaneously. Could be a weakness if over used or under used, so people don't bother to look could have setting so messages are taken off by date i.e. birthday notice can come off. A change of position should be on Intranet. Thomas 7 April 04

Thomas' comment on the instantaneous advantage of the intranet perhaps demonstrated the most significant benefit. We can distribute notices in many other ways but notices placed on the intranet by any staff member are instantaneously displayed on all the computers in the organisation. Thomas' point about the weakness being that it could be over used or underused is also very perceptive of the function that the news page performs. It is a live user-centred form of communication. Users either under use or overuse it, as opposed to a staff newsletter where the editor decides the appropriate amount of information. Yet this is an advantage of the intranet. It gives the users the means to establish the protocol.

Shirley's comment about how the photos make it personal has been mentioned before. This is a social benefit, as long as the photo is flattering. We did take care to take a number of pictures with a digital camera and ask staff which one they preferred. This personal emphasis of having pictures attached to messages was deliberate for this reason and is a striking characteristic of the intranet. When someone puts a message on it, it feels like it is from a real person. This perhaps explains why the social content such as referring to birthdays or someone wanting to sell a car or find a flatmate became part of the news items from the beginning, without any prompting. This mix of social and work communication reflects the culture of the organisation. Romus is a stressful place to work at times, with its fast pace of change, but it is also a friendly place, reflecting the personality and influence of the owner and president. The answers to the question below also reiterate these comments.

Q3 what is working on the Intranet and do you have any ideas about why this is?

> New people, leaving info get put on and that is important. Spin off of reorganising the I drive. Sally 21 Oct 03
>
> Calendar: People cannot be bothered clicking to see future only good as one form of reminder. E-mail: Rather use their own Outlook, which is better. Files: Not visible. Forum: Bug, could be a training issue. SMS: Have it as favourite on Explorer. Photos: Because we haven't added links i.e. preferred with favourite. Vaughn 11 Dec 03
>
> Most other things. People don't know how to use it or don't see its relevance. Shirley 1 March 04
>
> The need to disseminate accurate info quickly as a non paper (electronic) level. Thomas 7 April 04
>
> Newspage, of interest to staff. Contacts, needed. Google, guide access, but doesn't allow you to go back and forward. Jason 2 June 04
>
> General news and communication. 1) It is on the front 2) It has obvious benefits 3) It is easy to use 4) It hasn't changed much. People are familiar. Charles 2 June 04
>
> Don't use it sufficiently to comment much. I read/glance at general news then minimize it for rest of day. Douglas 1 June 04
>
> News, because people check it and sometimes contribute – the photos make it interesting and the topics have wide relevance eg staff meetings. Claire 1 June 04

In addition to the points already mentioned, there are some new points that are being made here. Sally mentions the spin-off of reorganising the shared network (I) drive. This is the shared drive for the organisation. She is aware that the reorganisation of the folders on the shared drive came about as a spin-off from the intranet development. Discussion has previously covered how the intranet uptake had to do with benefits compared with embodied world ways the organisation did things. The point is also being made here that intranet uptake is also affected by benefits of competing uses of other technological forms, such as the Windows interface and Windows tools such as Outlook Explorer for email. In the earlier iterations, a fast email feature was included on the front page of the intranet. But this was never used because staff already used Outlook Explorer and this had features which the intranet did not offer. Therefore, even if the intranet had offered a consolidated interface design, this was not beneficial enough to compete with dedicated email programmes. In such a case a user will opt to use one of the two ways to send email rather than send email through two different channels. The same issue applied to the Calendar option, which was removed in previous iterations for the same reason.

Q4 What is not working on the Intranet and do you have any ideas about why this is?

> Not computer literate. Vaughn 11 Dec 03
>
> They don't see a need for it. They don't know how to log in. They don't know how to use it. Shirley 1 March 04
>
> Don't have skills or no computer. Jane Liu 26 March 04
>
> Don't know. I do more likely to look for something rather than view news. Thomas 7 April 04
>
> Guarantee that everyone is aware of the communication. Not everyone has a computer and those that do, do not all view the Intranet. Could have an option to notify private email addresses eg if someone is sick. Thomas 7 April 04
>
> Fix it form. No training, possibly people do not understand or see the need. Manager's corner, too busy to update. Forums, no training like fix-it. Jason 2 June 04
>
> The filing page. People know windows and maybe more comfortable using it rather than using a new system that offers few advantages. Charles 2 June 04
>
> Only as good as the information placed on it. High reliance on humans/staff to maintain the information. Douglas 1 June 04
>
> Everything else except perhaps the contacts list people don't see why/how it will help their job. Claire 1 June 04

My philosophy was to make the interface design so user friendly that computer literacy is not an issue. All users had different levels of computer literacy but if all users were to use the intranet it needed to be designed to the lowest common denominator. I believe the comments related to training and computer literacy have more to do with the lack of perceived benefits of areas such as the forum. In fact there were training sessions on the forum, report area and file location feature, but they did not lead to uptake. In the case of the 'Contacts' page the comments regarding usage were more to do with the need for it to be updated in order for it to be effective. The difference was that in the case of the contacts page, staff had tried it because they could perceive the benefit, whereas with some of the other features they did not have a need to even try it. My interpretation, then, of the comments that training needed to be provided or that some staff were not computer literate enough was that it reflected more that there was no perceived benefit. This is the point that Claire makes. Douglas' comment that the news page was only as effective as the information placed on it and that it was reliant of humans addresses the issue of user-centeredness. This is exactly right, and shows that communication management is required not just to manage coexistence between embodied and virtual world practice but also to manage the communication process across the organisation.

Q5 why do you think some people never use the Intranet?

> Email, SMS, Links. Vaughn 11 Dec 03
>
> Yes, some people communicate via computer, some don't. Some people prefer face to face contact. Shirley 1 March 04
> Files not necessary. Jane Liu 26 March 04

Shirley made the point that some people will choose to communicate face-to-face. This is not such a significant point because even at an embodied world meeting, some people choose to not communicate much, if at all. Some staff prefer to communicate on a face-to-face or one-to-one basis. The intranet needs to be understood as only one mechanism of communication which coexists in with other virtual and embodied world methods.

Q6 What would you cut out of the Intranet?

> Nothing at this stage. Vaughn 11 Dec 03
>
> Links, Forum, Files. Warwick's not suggesting breaking up news into sections such as social and work based. People

> look at who puts stuff on presently senior staff. Shirley 1 March 04

Shirley's comment about people looking at who is putting messages on the news page demonstrates how a user-centred feature requires cultural permission within the organisation for user-centred activity to really happen. It reflects the issue previously discussed that non-assertive people in the organisation are not listened to properly. The point has been made that as an entrepreneurial organisation staff tend be listened too only if they are assertive. It is natural for staff to look at the intranet with the same understanding. It has been difficult to create a level of cultural permission in the organisation for lower level administrative staff to contribute as I had wished. To a degree, this permission can be encouraged by training all staff, so that they realise the intranet is for them to use as well as management. This issue also arises from context outside the organisation where the cultural expectation, in the experience of many employees, is that it may be better not to speak up. In this sense it is enough for the intranet to open up the organisation in the direction of participative collaborative communication. It also needs to be recognised that communication management needs to be resourced by the leadership of an organisation, and that that communicators and human resources staff have a common purpose and need to be technically savvy in order to manage an intranet resource. Such resourcing would reflect whether an organisations leadership recognised the economical benefit of collaborative communication.

Q7 What would you add to the Intranet?

> Suggest front page only because people are lazy and won't go to links page i.e. number of clicks same as going to Google. Shirley 1 March 04

It is much better to have as much functionality on a front page and that having to go down a level or tabbing to another section generally means that those sections will be less visited.

Q8 How would you continue the sentence "Our organisation is like a ...." Can you see any bearing between what you said and the development and implementation of the Intranet?

> Car engine. Intranet's trying to make engine go more smoothly like oil. Vaughn 11 Dec 03
>
> Family. No appointments – like a family. Like a family

barging in each other. Nobody knows how structure works. Jane Liu 26 March 04

Madhouse. Too many people disconnected from each other in their work, thinking they are actually doing what they are supposed to do to enhance the overall direction. Therefore a tool like the intranet is quite vital to help bring more cohesion. Jason 2 June 04

Family owned shop. People go to offices to tell people what they want rather than procedure – so we don't need to go through system therefore not useful. Jane Liu 26 March 04

Friend. We are a support and a partner to our staff. The Intranet as a partner and friend of the staff talks to the staff and we can all put into it. It is not from above – it includes support and advice. Thomas 7 April 04

River. Flow and course with change. Yes. In a relatively short time what seemed relevant with the Intranet quickly became in need of modification to meet new needs in the organisation. Frequent course changes. Charles 2 June 04

Series of semi-isolated villages, with intermittent communication between villages, at the whim of the local chiefs. Communication via the intranet is still at the whim and inclination of the "Chief" and acts as an inter-village notice board. Because it will never be possible to get people to contribute anything other than what they want to disclose, the intranet will never truly inform staff. Douglas 1 June 04

Lava lamp. No-one knows what it will do, it can move and change in all directions, some parts going fast, some going slow. Intranet is similar, and good to see it adapting to the style of the admin and management staff, (perhaps the fast bits) but not necessarily teaching staff. Claire 1 June 04

Jason's comment about the need for cohesion for the organisation perhaps clarifies the main benefit of the intranet for the organisation. From the comments above, it can be seen that the organisation is not culturally noted for its strategically structured approach. The reasons for the organisational descriptions above can be understood because of issues relating to the high rate of change in the organisation's internal and external environment, and also to the entrepreneurial nature of the organisation. In such an organisation, the question of how to bring about cohesion is a very challenging one, but it seems to be very much an issues. Strategies are warranted that will build cohesion and interaction. Perhaps the main positive outcome of the project is that the intranet helps build some cohesion by adding a communication channel, although its original aim was primarily opening up user-centred participation. On the other hand, it does seem to offer some user-centred participation as well, but not as strongly as I first hoped. The intranet itself is a cultural artefact and cannot be seen as isolated from the issues of communication management.

Claire also points out that the teaching staff is not necessarily included in the intranet communication channel. This is because they only have shared access to a computer and when they use it they are not concerned to communicate with staff across the organisation. In another department, the teachers each have a computer, yet they do not contribute either. They either do not see the need to contribute through the intranet, or, perhaps, may feel they have not been given the permission to use it. The intranet enables all staff to contribute directly, but it will not achieve this where cultural ways of doing things do not support a networked organisational paradigm. The points made below reiterate this.

Q9 "Organisational communication" is about the way we communicate at Romus. It's about how we communicate from the top down, between departments and with each other in general. Can you think of any communication issues that might have affected the development and implementation of the Intranet?

> Henderson didn't want to participate. The Franchise didn't quite feel a part maybe otherwise fine. Vaughn 11 Dec 03
>
> People feel like only managers and or organisers should put things on the Intranet, and also that they are the only ones who need to use it. Because managers created it, but didn't tell them it's for them. People haven't been told/insisted to use it. Shirley 1 March 04

> Habits are stuck in what they're used to with growth people kept doing same way no matter the size. Jane Liu 26 March 04
>
> Most people in the company are on our network (i.e. access to computer). Therefore it is logical to use a method. We don't meet in big group meetings every day or have a whole staff assembly every day - then that would be more suitable, but we don't. I still meet new people that have been working here for some time. Thomas 7 April 04

Vaughn's comment about the Henderson and the franchise schools reflects the embodied world business and communicative relations. The relationship issues with the franchise organisations have been explained previously. There is a direct correlation between embodied world relations and willingness for them to use the intranet. Thomas points to the fact that if more meetings happened that would be better than intranet communication. Where there simply is not the time to have meetings, the intranet is one effective communication channel to manage some aspects of communication. The cost of running meetings is also significant and in a busy environment, can be resented when too many meetings take away from needed work time.

Q10 "Organisational culture" is about the way things happen at "Romus" as compared to other ways companies can work. It's about management styles, cultural different ways of doing things and personality mixes. Can you think of any cultural factors that may have affected the development and implementation of the Intranet?

> Forum thing. People don't want to put too much in writing – could be because of taboo subjects i.e. there might be an agenda they're not aware of so what they say might come back on them. People do things behind closed doors, so policies not transparent agreements and what people are told may not be upheld and can be changed at whim. This lessens the buy in for open discussion because tomorrow could bring a totally different policy. People put time and effort in to solutions, but others could be working differently on the same solution. This is why people won't get hooked on the forum. Sally 21 Oct 03
>
> Communication of events and contact details and why they are. Vaughn 11 Dec 03

> Contacts: Direct dial, extension and little email link.
> Shirley 1 March 04
>
> Yes. Notice board is good – nothing else to replace. Jane Liu 26 March 04

Sally makes a strong statement about the organisation's culture and why she does not expect there to be up-take of the forum. Her comments reflected disenchantment with the way she believes agendas work in the organisation and in fact she handed in her resignation not long after making this comment.

While an organisation with an entrepreneurial culture will undoubtedly have a degree of perpetual disorganisation, it is also a model suited to a networked organisation working to survive in today's market place. The organisation has developed through a series of market declines and dramatic changes. In such an organisation, I believe staff should not ask it to be different, rather should decide how to contribute to its efficiency, and in so doing, find an appropriate cultural role, and can feel a degree of belonging. In this organisation, keeping a sense of humour is essential. The culture of the organisation is not supportive on the whole of participative and collaborative communication. Staff opinions are warmly received through their assertive involvement. In such an environment the need for cohesion perhaps is of primary importance if significantly greater participative collaboration is to be achieved.

Q11 Do you see the Intranet offering any better ways of interacting than exists otherwise?

> Booking system would be good. Social bits could be expanded. Sally 21 Oct 03
>
> No. Vaughn 11 Dec 03

Q12 How could I as the developer and implementer have done a better job?

> More training and encouraging people to use it face-to-face. What's missing is training. Must have face-to-face explanation. Better if it can be used as a communication channel, but ESOL management also needs to take responsibility. Issue is organisation is that people don't relay information to relevant people, without being instructed – same with info that goes on the intranet.
> Shirley 1 March 04 IT

> More training would allow us to explore parts that we don't use. Thomas 7 April 04

It was a dilemma for me to manage the intranet project within a hectic work environment. I was not convinced that a lack of training had undercut the project. During the stages of iterations, it became evident that the benefits were not significant compared to either other embodied or technical ways of achieving the same thing. Responsive designers can decide to alter the design when user testing makes obvious that one aspect has no benefit. Everything that happens within the organisation happens in a hectic environment. More training would have helped, but the reality is that this was not possible. Yet the organisation does adapt to new procedures. The challenge for me as a developer was to make the intranet so user friendly, with immediate benefits, that uptake would happen with as little prodding as possible.

> Try to find out more what people need. Eg file system not necessary. If everyone has a phone list they don't need it duplicated on the Intranet. Jane Liu 26 March 04
>
> More research up front of what staff wanted/needed. More training. Imparting vision to staff and management. Jason 2 June 04
>
> More preparation of staff and management prior to development. Provide examples of existing systems. Charles 2 June 04
>
> Ask people what they want out of the Intranet, what they would find useful – preferably in a face to face discussion group type setting. Claire 1 June 04
>
> Generally design seems to be ok. Training sessions at management level and introduction of Romus systems that required mandatory frequent use by managers might have helped. Douglas 1 June 04

From the designer's point of view it was clear during the early iterations that users did not comprehend what could be achieved through the intranet and that therefore, the process of development required a series of iterative development cycles. As the project developed, users became more able to critique the intranet. As the project also involved exploration of the cultural, communicative and knowledge aspects of the organisation, the development story was an evolutionary process. In this sense, Charles' suggestion to prepare staff better was not possible. The dilemma of undertaking the project on top of my management responsibilities meant I did the best job I could, with the resources available. It had the advantage of gaining an insider's point of view.

Q13 As the developer what values, biases assumptions or goals do you think I have had?

> Probable assumptions that it would be much easier to get people buy-in, and folk would understand the reasons/visions behind it that Romus could easily be swayed – big mistake! Jason 2 June 04

> Understand the culture and its resulting characteristics, you appear to attempt to improve communication and clarify boundaries and from these two areas many of our problems stem. Charles 2 June 04

> 1. Assumption that all people value communication. 2. Did not distinguish between communal news and management communication when implementing system. Douglas 1 June 04

This assumption about my beliefs and agenda are correct. These comments show that the influence of the literature review did strategically influence the direction of the design.

> I don't think that the forum will be used much because people are too busy and don't see the relevance. This is an area that lack of training and/or explanation of what it is why we are doing it, is evident. People will not likely use something that no one has talked to them, personally about. Claire 1 June 04

In fact there were two training sessions to try and get staff to use the forum.

Q14 Another person that I interviewed said that.... do you agree with that?

Sally's points about agenda – no don't agree because news isn't political and poll is anonymous. Vaughn 11 Dec 03

Sally's comment on decisions being made behind closed doors – in reality decisions are made sporadically – people who make decisions don't think about who's following and so they won't notice when people are off track. Thomas, will assume you know – he talks about something you know nothing about. Classic: Janice waited for him to come around to the subject again and asked for details. Whereas if you get stressed by his approach it doesn't work with him. Interest is sporadic, when interest starts changing and previous decisions are forgotten. There is an entrepreneurial disorganised – shooting from the hip approach – there is initial interest, but steady progress is not of interest. Shirley 1 March 04

As per Sally. Yes I agree with Sally's comment on decisions being made behind closed doors – but depends - it's about nothing getting done. Speaking up doesn't work because people already have their mind set on certain decisions and priorities. It's all about what Thomas wants – either you try and change his mind or your opinion isn't worth much. This is why no one has used the suggestion box. There is no process or person for Thomas to deal with things. SMS works because there is no other way. If Intranet contacts are not up to date they'll use the other phone list. Note: Sally was told to keep the non Internet phone list updated, therefore contradictory departments. Jane Liu 26 March 04

Thomas's point that the intranet communicates in ways the organisation does not favour in real life, preferring a daily meeting, aroused disagreement in some staff members.

The Intranet is not just a news page. Even if meetings are much better at communicating news – meetings shouldn't focus on news dissemination. The intranet is a communicative tool on potentially many diverse levels, recording processes, exchanges, input, much better than

> this perception – which underlines my belief that the intranet lacks Thomas's backing in any substance – not by ill-will but simply by not grasping its value and place in the organisation. Jason 2 June 04
>
> Putting thoughts in writing, in public is a method that offers advantages that oral presentations at meetings do not. Public record will help clarify things for people, inform the absent and document details for future reference. Charles 2 June 04

On the other hand, some staff members agreed with Thomas's point of view:

> Agree but only works for those that regularly log on. Many staff do not. Douglas 1 June 04
>
> A lot of the news is things that are quite immediate, so would be forgotten at a weekly staff meeting, or are things that are not appropriate for a staff meeting. Eg "I lost my wallet". Therefore I think that the 'news' would still be useful even if we had weekly staff meetings. Claire 1 June 04

Apart from the variety of views expressed here, the entrepreneurial nature of the organisation is emphasised. Opinion was expressed that it would have been preferable to have had more support for the intranet from the leader of the organisation. However, from my point of view these comments illustrate common organisational frustrations. Any newly implemented system will need to overcome a lack of appreciation for its purpose. Not only does it take time for organisations to change but also, intended change rarely happens as desired. The intranet has been adopted in an evolved form which represents user choice.

Q15 Is there anything else you would like to add?

> I believe in the idea, so I think it should be continued. Jason 2 June 04
>
> The concept is good, especially in an organisation such as ours. The real benefit I expect will only come in conjunction with other developments over a longer period

> *of time. Charles 2 June 04*
>
> *We need to figure out a way to include more people in the design and implementation. Claire 1 June 04*

The second set of data was gleaned from interviews conducted on 6 June, 2004. As discussed in the previous chapter, by this time I believed staff were familiar enough with what I was trying to do with the intranet and management interventions that they would be able to comment more specifically in relation to the thesis propositions. The section headings and italicised text introduces the propositional context.

Influence of the wider business context

Romus exists in a global knowledge based economy, where issues such as exchange rate, international competition and our entrepreneurial culture mean we have a high rate of change and adaptation. Characteristics found in companies operating in this economy include organic, cooperative, participatory, quality led and customer driven.

Q1 Do you think any of these characteristics apply to Romus and/or what characteristics would you attribute to Romus?

> Romus is organic like a jungle, cooperative like chewing gum, participatory for folk with initiative, idea but not necessarily quality led, and again, more driven by ideas of management than proper market analysis. Jason 2 June 04
>
> Yes. Romus has a liquid element, whereby it can encompass changes and new opportunities almost without staff realising the changes happen. Charles 2 June 04
>
> Romus has high rate of change of direction – not so much global economy. With low tariffs etc all NZ businesses are exposed to international competition. Douglas 1 June 04
>
> Yes, organic, cooperative, participatory; perhaps not really quality led in terms of lack of follow up on customer comments/problems. Or customer driven to be frank. Claire 1 June 04

Does the organisation fit the networked organisation paradigm? According to these comments, it does.

As a theoretical response to the influence of the wider business context, the intranet has attempted to be participatory - to get staff and management to provide content. The News, Calendar and Contacts pages have been designed for user input.

Q2 Do you think this participatory emphasis is a good idea and why or why not?

> It is good only to the extent that it achieves user participation. For which more thorough training to users is required. This has not yet been accomplished at Romus, and therefore user participation is fairly minimal. Jason 2 June 04

> Yes. The way and speed in which changes occur mean that communication between staff is essential if everyone is to be involved in moving forward together. Charles 2 June 04

> Participation is at the less critical level – intranet does not communicate much of the wider business context' bigger issues. Douglas 1 June 04

Yes, it's a good idea but I worry about whether the end user, or at least some of them, were considered in the design and implementation eg it does not really encourage participation by teachers (half of our staff). Due to its design but also due to lack of training and lack of application of limited computer knowledge and skills of some staff. Claire 1 June 04

These staff affirm the idea of increased staff participation. The intranet can only be one mechanism for change in this direction. There have been many increments of direct intervention by company management to achieve a greater participative workplace, and this need is ongoing.

Culture of Romus

Organisational culture is about the way things happen at "Romus" as compared to other ways companies can work. It's about management styles, ethnic cultural different ways of doing things and personality mixes. It is about the unwritten rules.

Q3 Can you think of any cultural factors that may have affected the development and implementation of the Intranet?

> One of the biggest reasons why the intranet has been faltering so far is the lack of a consistent culture from the top. Superficial buy-in but no commitment to focus on intranet as a design tool to form/normalize Romus culture. Jason 2 June 04
>
> Yes. A lack of clear direction from different levels of management meant different levels of buy in existed in different departments. Charles 2 June 04
>
> Decisions (the way things happen) occur without proper consultation and process. The internet has not altered this culture. Douglas 1 June 04
>
> Yes, a bit of an in-club that knows what's going on, and that changes things quickly, without much communication to those on the boarders or outside. But, on the other hand, I think the Intranet has helped heaps to lessen the impact of that; aiming at top management being assisted in communicating to those below. Claire 1 June 04

I believe Claire's comment to be particularly insightful, in that because, the organisation is entrepreneurial and has a high rate of change, a mechanism that increases participation and communication provides a cohesive influence and this organisation needs cohesion.

The Intranet development and implementation process has involved some management interventions by Warwick. He has been trying to get participative involvement and to use the Intranet to bring cohesion in areas such as communications and knowledge management (manuals). Attempts at getting reporting to work through the Intranet have been made.

Q4 Do you think either the Intranet itself and/or Warwick's interventions have affected the way things are done at Romus (The Romus culture)?

> Both a definite yes, though at times it seems like fighting a losing battle, given the highly diverse and entrepreneurial nature of the beast. Jason 2 June 04
>
> It hasn't changed the culture yet. It has changed/begun to change the way things are being done, particularly in regard to communication. In future, these changes may result in changes to the company culture. Charles 2 June 04
>
> The culture hasn't really changed, but I think the attempts at the Intranet have highlighted the difficulties with the present culture and its resulting lack of communication. Claire 1 June 04

I would not expect to change the organisation except by a process of influence. I am not the chief leader of the organisation and so I do not set the primary agenda or culture of the organisation. However, I am the manager in the organisation who brings a level of strategic and communicative cohesion. This is a key value of my role as Principal. It is realistic that managers in typical organisations can bring their own influence to bear, but the nature of working together means that culture is a negotiated dynamic.

Communication systems of Romus

It has been said (Roger D'Aprix, 1999) that an organisation's communication process reflects the leadership of that organisation. Communication processes can include any form of meeting, dissemination of information, way of handling issues, mechanisms of management and staff to communicate to each other. It is about the way we communicate from the top down, between departments and with each other in general.

Q5 How does Romus communication process reflect its leadership?

> Very inconsistent, sporadic, and based on a buddy system where being on the inside counts – rather than a thought out communication process on a need-to-know basis – typical for entrepreneurial organisations, but very unhealthy if continued after initial start up. Jason 2 June 04

> Sporadic, occasionally brilliant, generally lacking unity and usually in varying states of disrepair. Charles 2 June 04

> Romus communication is very variable in quality. Effective management requires frequent communication between the leaders, not general communication to all staff via the intranet. Douglas 1 June 04

> Haphazardly, with the result that some feel included, others excluded. Hard to keep up with changing leadership direction, biases, preferences, motives. Claire 1 June 04

There is a definite feeling that the communication at Romus leads much to be desired and that the entrepreneurial culture is partly to blame. In a Quality Improvements day held in February 2005, I asked all the managers and administrative staff to draw a network diagram representing each person's informal working networks. Following this we posted the diagrams on a wall for viewing and invited discussion on issues. The discussion of the diagrams established the reason for the networks and two key issues. First, as with most organisations, the informal networks existed so that people could get their jobs done. The discussion showed, however, that there should be a balance between line management protocols and informal working networks. If there was too much protocol, action got bogged down in red tape and tasks became too difficult to accomplish. If there was too much informal networking, however, decisions were made without proper involvement of key personnel, and everything became too disorganised. This last point, that of failing to involve the right people, was seen as a significant issue, especially in one department, where the teachers felt excluded from the culture. Discussed revealed that strong feelings existed that management had not supported teaching staff in the right way during the previous year's boom expansion. Some resolution was reached in discussion.

The network diagrams showed how different individuals approached networking. Some staff networked in a way which reflected clear line management and protocol emphasis, while others networked in a way that showed no connection to appropriate line management. From my perspective, the diagrams clearly reflected where the major internal working issues exist in the organisation.

My objective for this training session was to raise awareness of communication, legitimise the networking propensities of the staff and to engender frank discussion of the issues involved in working through informal networks. At the beginning of the session, I introduced the concept of networked organisations and used this as a rationale for the permission I offered them to continue to work in a networked manner, with some restraints for the sake of organisational function. I hoped to encourage cooperation and personal responsibility about the way staff established their informal networks. As Principal of the organisation, I have embraced the networked organisational paradigm and will influence the organisation in that direction. With continued support from management, the potential benefit of a user-centred intranet is more likely to be realised. I do not see this as an excessive use of authority, but rather, as a legitimate management intervention aimed at suiting the organisation's style of activity to the internal and external environments in which it operates. The point that I was able to make to staff about their style of informal networking proved to be, for me, a positive spin-off from the intranet project.

Q6 What ways do we communicate that might have impacted on or been influenced by the Intranet development and implementation?

> Official staff announcement made now via the intranet (though a number of staff (i.e. teachers/ESOL) will not get info this way. So far sadly not much else. Jason 2 June 04
>
> The fact that different departments can operate in isolation for lengths of time hinders development of a centralised system, particularly without managerial support of such a system. At this time it feels like there is a real struggle within the organisation between managerial culture and the common sense of improved central communication. Charles 2 June 04
>
> Intranet has informed the masses but not assisted management decisions. Douglas 1 June 04
>
> The problem is that things happen so fast, and feelings and biases are hard to reflect on the intranet. The result is that the most important events and/or directions are not really covered on the Intranet, at present. Claire 1 June 04

Knowledge management of Romus

There are two kinds of knowledge. The first kind is the hard kind that ends up being kept somewhere. It includes any policies and procedures, SMS, reports, meeting minutes etc. The other kind is the soft knowledge that staff have about their jobs, which can often not be captured, recorded or passed on to new staff.

Q7 Do you see the Intranet as having improved our management of hard knowledge in any way?

> Sadly, not yet. Though it has the seeds of possibility, and various attempts were made. Crucial – the area of training. To my mind that has been lacking – plus hardware access for some staff teachers. Jason 2 June 04

> Yes. Promoting developments and improvements in the storage and access to hard knowledge has been successful. People know where to look when something can't be found or has been moved. Charles 2 June 04

> Not really, but I am very computer literate so would not need the internet to assist me in finding hard knowledge (except phone numbers). Douglas 1 June 04

> Yes, gives an outlet to post up info about say, changes to staff handbook, changes in birthday policy etc. Have found it very useful in this regard. Claire 1 June 04

Q8 Do you see the Intranet as having improved our management of soft knowledge in any way?

> In small measure, especially as with Q9 above, ongoing events. Again, training is missing, and the top level support/endorsement/political will to make it actually useful. Jason 2 June 04

> Beginning to. The lack of detailed awareness of boundaries means that much of this soft knowledge does not lay with the source people expect hence the cracks through which the shit falls. Charles 2 June 04

> Don't know of any area of intranet relating to soft

> knowledge. Douglas 1 June 04
> No, not really. Claire 1 June 04

Potentially it was the forum that could have captured soft knowledge. However, as previously discussed the benefit for staff to use the forum was not significant enough for uptake. Staff could network in the organisation to achieve discussion on issues significant to them. The benefit of the forum would be one realised by management in saving a history of discussion and solutions. A one-sided benefit like this was not going to be enough a reason for staff to want to use the intranet.

2.4 Conclusion

This chapter has told the story of the revamping period. It outlined the management interventions that took place during this time and how developing a council reporting system, monthly full staff meetings, and revising the manual files were effectively spin-offs related to the intranet development process and thesis exploration. I concluded that knowledge management and reporting methods needed to change to suit a hectic, ever-changing internal environment in which restructuring is frequent, and I saw that key resources, including time, might not be available for these important elements of organisational life. I believe that this situation may not be the case in every organisation, but it certainly is at Romus, where constant change and restructuring consume human and other resources.

This chapter explained how the knowledge management feature of providing a menu system for access to shared files was competing against the windows interface staff were used to and which had additional advantages that outweighed the advantage of the new menu system. This comparison between competing systems showed how the benefits of this feature of the intranet were negligible. Nevertheless a spin-off was the re-sorting of files on the shared drive.

We explored the revamp issues. Staff comments make their views of the organisation clear, and reveal the cultural and communicative issues affecting the intranet development. These comments have shown that the communicative function of the intranet is the most significant for them. This communicative function is not so radical that it can transform an organisation, but it provides a new communication focus that influences the culture of the organisation. Other spin-off benefits occurred because of the project. These spin-offs were the creation of a monthly staff meeting, which included participation of a wide range of staff, and the council meeting, which was a collaborative way of managing reporting to senior management.

Comparisons between embodied world and virtual world mechanisms had been a feature of the design and exploratory processes. The communicative and cultural benefits of the intranet were now articulated by staff, in such a way that increasingly addressed the thesis propositions.

SECTION THREE

THE DEGREE OF INFLUENCE

3.1 Introduction

The intranet project has involved developing and implementing an intranet in the educational organisation where I work as the Principal. My role in the organisation meant I had significant leeway in the degree of experimentation both with the intranet and with management interventions to support the project. The situation afforded an opportunity for the research project, which provided a robust testing ground and exploratory environment for the thesis propositions.

My intention was to develop an intranet according to a user-centred design and to test the design and implementation against the thesis propositions. The first thing I wanted to establish was that a user-centred emphasis in intranet design is necessary in today's knowledge based global economy, where there is a high rate of change. The second thing I wanted to establish was that a user-centred intranet is not only an organisational artefact, but also the process of developing and implementing it will open up an organisation's culture, communication, and knowledge management in a participatory way.

The user-centred approach was intended to coexist with a participative and collaborative workplace. The paradigm for this kind of workplace is a networked organisation. I not only embraced the paradigm of the networked organisation, I supported and sometimes adopted a communications management role.

A user-centred intranet is conceived, primarily, as a device to encourage organisational cohesion by providing communication opportunities for organisations with a high rate of change in the modern knowledge economy. An intranet is a mechanism for inter-organisational communication and information transfer, and this is, perhaps, its main value. In large organisations, particularly, there may be many benefits for staff dispersed by distance and geography. Traditionally, the focus of intranets has been to disseminate human resource communications, and design that considers and collaborates with users is significantly different from old top-down dissemination of information.

The project has specified the benefits of technological mechanisms by comparing embodied world ways of doing things. For example, email has the benefit of discussing things to and fro. Doing this by email takes less time than organising meetings. However, when matters are complex, real meetings are better than email because there is more responsive opportunity and more embodied world clues to assist the communication. An intranet is one type of technological mechanism, and so when trying to realise the benefits, its potential use needs to be compared to embodied world ways of doing things as well as other technological mechanisms such as email. An intranet can have features which provide access to an organisation's manuals, policies, procedures and forms. However, these can also be accessed via the standard Windows interface in a shared internal network, so it is likely that staff who use these items will compare the merits of the two technological mechanisms and decide which gives best access for them. Designing a user-centred intranet means offering features which provide the opportunity for capturing the know-how and creative participation of staff, but it is clear from the feedback that I received that staff need to perceive the benefits of contributing in this way.

In a knowledge-based and globalised economy, the rate of organisational change is significantly increased from the industrial economy model. A networked organisational paradigm provides an appropriate paradigm and vision for organisational architecture. A user-centred intranet is suited to a networked organisation, which is necessarily communication rich, valuing collaboration and participation. Many organisations are evolving towards a networked organisational paradigm, but there is no single "right" model for a network. Obviously, different organisations develop different forms of structure and culture that enable different degrees of collaboration and participation.

A user-centred intranet may be an opportunity to open up the culture of an organisation for more collaboration and participation. It is not possible to attempt to implement a user-centred intranet without simultaneous supportive intervention from management.

This intervention expresses itself in the architecture of the way things are done in the organisation and also defines the role of a user-centred intranet in the boundaries of Human Resources and IT. In this sense, the designer of a user-centred intranet will simultaneously aware both of the technologically involved in the design and also of real use made of the technology in the embodied world. This develops a communication for the designer that bridges the "agenda divide" between IT's focus on security on the one hand, and the creative marketing focus of New Media on the other.

When an organisation begins to develop a user-centred intranet, the process may bring to light issues connected with the degree of genuine support a leader is prepared to give to this type of networking in the organisation. It also brings tends to show the attitudes employees hold towards the organisation and to using new technology. In a networked organisation, staff need to balance their adherence to established protocols of line management and to agreed guidelines with self-designed informal work networks. The balance is necessary, because imbalance leads to either degrees of disorganisation or inflexibility. A user-centred intranet opens up an organisation's communicative culture by the process of implementing it. It brings into focus how staff in an organisation communicate and what cultural permissions exist.

Internet technology has been evolving towards a user-centred paradigm. These same technologies are used for an intranet. The interactive communicative functionality is technically enabled by the emergence of online html to database interactivity. This is governed by the *.asp file convention. For example, instead of an administrator downloading staff pictures, the software can be adapted so that users can upload the pictures themselves. Instead of newsletters and messages being solely created from management, all users can be given the facility to post messages.

The user-centred and interactive technology designed into an intranet can encourage cohesion, a sense of belonging, and improved communication, and the benefits of its communicative features seem to outweigh other features such as storing explicit knowledge in the form of manuals or reporting features. This does not mean these features will not benefit the organisation, especially larger and geographically dispersed organisations, but no matter the size or geographical placement of the organisation, the communicative element is likely to be of the greatest benefit.

Implementing a user-centred intranet might be a strategy for business survival and competitiveness, because it could support the cohesiveness of the workforce. It may be one tool that helps secure an organisations existence in an ever-changing business environment.

3.2 The organisation in a knowledge based and global economy

The first idea I wanted to explore, that a user-centred emphasis in intranet design is necessary in today's knowledge based global economy where there is a high rate of change, required an exploration of how Romus fits within such an economy. It suggested the necessity of exploring the concept of user-centeredness.

The response of Romus to its international educational markets is to continually adapt. The owner and driving force for the organisation is an entrepreneur who drives change in order to capture business opportunities. The culture of the organisation adapts through restructuring, role changes, and informal working networks. The rate of change, together with the leadership style of the owner, can sometimes create a state of semi-disorganisation.

As an educational institute serving international students, the organisation is very much connected to the global economy The business of international education is influenced by international politics, exchange rate, place in the global market, economical situation of the overseas market, and shifting educational perspectives of the overseas markets.

The organisation responds to this in an entrepreneurial way, both by developing its products and also by aligning itself with strategic partners and in some cases inviting business partners to be involved within the organisation.

Descriptions gleaned from interview data referred to the organisation as: Family, Madhouse, Friend, River, Lava lamp, and a series of semi-isolated villages. Staff used terms such as, organic like a jungle, having a liquid element, a high rate of change of direction, cooperative, and participatory. The organisation is certainly not structured in a static way. The challenge of maintaining cohesion in such an environment is a need that clearly exists for Romus Colleges. It is a modern organisation ,which has a high rate of change.

In the previous chapter, data from the interview process reiterated that Romus Colleges was, albeit imperfectly, characteristic of a company operating in a knowledge based and global economy. Jason referred to Romus as "organic like a jungle, cooperative like chewing gum, participatory for folk with initiative, idea but not necessarily quality led, and more driven by ideas of management than proper market analysis". June emphasised the rate of change when she said "Romus has a liquid element, whereby it can encompass changes and new opportunities almost without staff realising the changes happening". Claire said that Romus was "organic, cooperative, and participatory". She also reiterated Jason's point about the lack of a quality led environment in saying that Romus was "perhaps not really quality led in terms of lack of follow up on customer comments/problems, or customer driven".

Senior managers in the organisation are also entrepreneurs. This can lead to activity which is not strategically aligned. Nevertheless the sometimes sporadic entrepreneurial culture of Romus Colleges brings an enthusiastic impetus. The entrepreneurial spirit of the owner and senior managers provides a type of cohesion of belonging for those in the organisation as long as what the senior managers are doing is passed on to staff. The intranet as a communicative mechanism added a communication channel for senior management to update staff on initiatives and developments, providing a level of cohesion.

One challenge for the organisation is to provide for the organic change, while maintaining a minimal level of structured protocols. If there are too many structured protocols participative collaboration is stifled and if there are too many self designed organic working networks then there is too much disorganisation. Cohesion is needed in this environment and the intranet has served as one mechanism for cohesion. It is a new channel of communication. The revamped intranet at Romus became communication focussed and being user-centred it invites collaboration by staff sharing what is going on, information on who is doing what and how to contact them, and pictures of people doing what they do around the organisation. The process of developing it also provided cohesion because of spin-offs. An attempt to create a reporting system through the intranet failed, but led to a new collaborative reporting meeting. An attempt to create a menu and sub menu system to access manual files failed, but the work to improve the files themselves created a quality improvement and more staff involved in updating files. An attempt to create a forum failed, but led to a monthly staff meeting which involved a wide range of staff sharing about their work was a spin off benefit.

Networked organisations

The networked organisational paradigm suits today's knowledge based global economy, where there is a high rate of change. Embracing a networked organisation facilitates structure that is collaborative and participative, which is essential when dealing with continuous change. The intranet project acted as a vehicle for change towards a networked organisational paradigm.

Undertaking the development and implementation of an intranet has been challenged in a sometimes hectic environment. Embracing a networked organisational paradigm provided a constructive way forward for introducing and implementing required changes. In such a paradigm, communication management becomes a feature of cohesion as does any mechanism to support networking and collaborative communication. The intranet evolved to supplement the networked organisational paradigm. The design process itself was user-centred, with evolving iterations being undertaken primarily following response from users.

The first idea is affirmed. A 'user-centred' emphasis in Intranet design is necessary in today's knowledge based global economy, where there is a high rate of change. A networked organisational paradigm is a constructive way to structure for a high rate of change. The characteristics of a networked organisation are similar to the user-centred design paradigm and the process of designing and implementing the intranet provided some cohesiveness in the move toward the paradigm.

Technical issues

A networked organisation may network across geographical boundaries. This raises technical issues. The intranet development and implementation process at Romus has experimented with the server location of the intranet and its components. If the intranet is located on a server outside of the organisation it provides the greatest opportunity for flexibility of use. It can be logged on to by divisions of the organisations in different geographical locations, sales and marketing personnel on the road or on international business trips, and from home.

However, this impacts on how an organisation uses traffic heavy data such as pictures, multi-media, and manual files. This project explored an auto-conversion process for changing the organisation's manual files into HTML and moving them onto an external server. In this way staff could update manual files and they became instantaneously accessible by all staff as a read only HTML version. In order for this system to be useful the files themselves needed to be useful. This required analysing the manual filing system and attempting to cull duplicate files, create file saving and naming conventions. However, the staff had different degrees of technical ability, and controlling file naming and saving protocols proved to be unreliable. Some staff developed files on their personal rather than the shared network drive. This in itself created duplicate files. The habit of emailing files o others also leads to duplication of files. Keeping filing conventions consistent would have required centralised control of manual updates and this would have defeated the idea of involving staff in updating their own manuals. If it was ever going to possible to maintain filing conventions a great deal of training and support would be required and Romus Colleges did not have the resources to do so.

3.3 The effect of a user-centred intranet on organisational communication

The second idea that I wanted to explore was that:

> A "user-centred" Intranet will not only act as an artefact, but also the process of developing and implementing it will open up an organisation's culture, communication, and knowledge management in a participatory way.

Having a user-centred intranet means that participative and collaborative working networks are supported with an appropriate communication channel. This is necessary in a networked organisation. Having a user-centred intranet, however, does not guarantee that participative and collaborative working networks are effectively created and managed. This, then, is the subject of this section.

I have already asserted that a user-centred intranet supports and embraces a networked organisation by facilitating structure that is collaborative and participative, thereby becoming a vehicle for change. It does this most predominantly in the area of communication, but to achieve effective change, it became clear during the project that, collaborative and participative structure needed to be created and managed by a manager of internal communications. The second idea I wanted to explore is affirmed, but on condition the process is managed. A communications manager, who embraces participative and collaborative communication, can be responsible for driving the real change. The intranet supports this change!

Communication success

The two most successful parts of the intranet were the news page and the contacts page. The news page has been very successful in terms of participation of users across the organisation contributing many kinds of information. The way pictures are attached to news items provides a personal feeling. The news page has been used for sharing both formal and informal information across the organisation, effectively becoming a main notice board for the organisation and enabling an increase in information being dispersed. Compared with other methods of communication, it has the added advantage that messages can be instantaneously posted on the computer screens of all users in the organisation. The contacts page keeps track of changing staff roles and responsibilities and how staff can be contacted. Although initially designed for managers themselves to update, it has become a process largely taken on by the communications manager. These pages talk about what is going on in the organisation, who is doing what, and how can they be contacted. As a communication channel the intranet does this significantly better than any other way. The staff in the organisation needs an easy way to know these things. It provides cohesion. Because pictures are used in both sections it works in a very personal way, and so it brings a feeling of belonging. This is another benefit of the system. Once this theme of communication became apparent some additional changes were made after the final revamp covered in the previous chapter. There were a special message section for the senior managers and the Fix-it section was replaced with student interviews, so that staff can see what is happening for its clients.

Here is a screen shot taken on 21/06/2005 which shows these additional features.

Figure 3.3.1 Additional features post revamp stage.

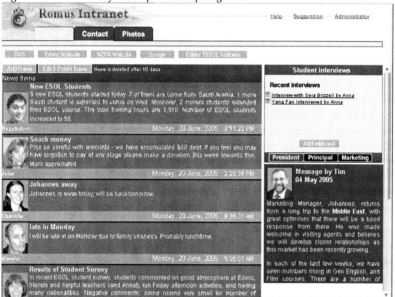

Other sections of the intranet were taken out because they were not effective. These included the file menu and sub menu system, the forum section, the calendar, and the reporting systems. When introducing technology it always has to be asked whether there is a significant benefit over other existing ways of doing the same thing. This is the reason these features did not work on the intranet. In the case of the file menu system the windows interface offered better versatility. In the case of the forum system, staff preferred to talk face-to-face. In the case of the calendar people prefer diaries that they can carry with them, in the case of the reporting systems managers in the first place prefer to control how they report and in the second place reporting face to face offers the opportunity to question. However, with most of these sections, management involvement to support the attempt to get them to work created spin off benefits. In the case of the forum we created a new monthly full staff meetings where staff were offered the opportunity to talk about what they did. In the case of the Menu system we revamped the manuals files for the organisation and set in place some filing conventions and protocols. In the case of the report section, a new reporting system was devised where reporting was managed in a collaborative way.

The intranet itself did not open up the organisations culture and communication, but the process of implementing a user-centre intranet did by means of the spin off benefits. By open up I mean raising an awareness of these areas and providing a means and driven process by which an organisation will change towards an effective model for business success.

The issue of a lack of strategic communication from the top managers was discussed. Charles commented on "the common sense of improved central communication" with use of the intranet, and Claire reiterated other similar comments that "the attempts at the intranet have highlighted the difficulties with the present culture and its resulting lack of communication". She felt the intranet had had a positive counteracting influence in saying "I think the intranet has helped heaps to lessen the impact of that [lack of management-led communication] aiming at top management being assisted in communicating to those below". This illustrates that the process of developing and implementing the intranet assisted in opening up the organisation's communication.

Access to the intranet

The Intranet needed to be set up and running on each computer and any login schemes needed to be working simply so as not to provoke misuse. As long as staff have primary access to a computer, then it is relatively easy to invite them to use it. Staff in an organisation with secondary access to a computer, however, are not as easily able to contribute. Because of the user-centred approach we were looking closely at the participation of staff, which would require all staff having access to the intranet. It was found that the teachers in the organisation did not have the same access to computers as other staff did, but shared secondary access to computers and therefore were somewhat disconnected from the intranet as a communication channel. The dual solution, which was pursued, was firstly to ensure all staff had at least shared access to computers. I also decided to recruit key staff in those areas where staff shared access to computers, to be key supporters for enhancing communications. However, this did not overcome the limits of secondary access to computers, and so it was determined that other channels of communication were needed for staff without primary access to computers.

This issue illustrates the fact that there are many channels of communication in an organisation. Among them are intranets, email, notice boards, informal and formal meetings, internal newsletters. It also illustrates that an intranet is a mechanism not a management tool. It enables collaborative communication, but a person, a communications manager needs to manage all the communication channels in an organisation providing for overall cohesion and coexistence between real world and virtual modes of communication.

Technical emphasis

Technical aspects associated with the intranet affected the comfort level of staff using it. For example, we needed to change the auto-delete feature on the news page for news which was more than 10 day old. It was felt that a 'use-by' date on news was needed as staff were not comfortable with having their old news items 'in people's faces', once the items had ceased to be of interest. This action was about trying to instil a positive feel with the way the news was handled.

This illustrates the need for a communications manager to be aware of how staff find and use all channels of communication. This naturally includes the technologically driven mediums and the need for a communications manager to be involved with new media designers of organisational intranets. Along the lines of the usability testing sessions run for the development project a communications manager needs to drive critique of any system.

Training and maintaining

In today's workplace communication systems such as email require staff to be trained in the medium. Since a user-centred intranet requires more than reading top down disseminated information, users need to be able to log on and participate by entering data. This requires training. Training also included raising awareness of what people in the organisation wanted to know from each other, because the training is for collaboration, not just how to enter data.

For the intranet to continue working effectively, it must be maintained and supported, because a user-centred intranet empties of content if staff do not generate it. A communications manager needs to ensure that information is generated and up to date. For example, attention had to be paid to the intranet after the Christmas break. In other cases we noticed certain departments were not contributing at all. This is also a good test of user-centeredness. If it is dynamic and staff stop entering data it will begin to look empty, like an empty room. On the other hand, if it is staff are enthusiastically entering data in a collaborative way it will look busy, like a busy room.

3.4 The effect of a user-centred intranet on organisational culture

Organisational culture arises from the way people interact. It is the way things are done and the shared beliefs and assumptions that exist. During the development and implementation of the intranet, there has been little significant difference to the participatory nature of the organisation's culture. It has, however, provided a degree of cohesiveness. It has been encouraging to see from the taped diary dated 26 September, 2002 that there has been a perceived feeling of belonging and pride in relation to "the intranet, how it gave people who were using it a feeling of togetherness".

In describing the organisation, interviewees referred to it as a family in which no appointments were necessary, a madhouse in which too many people were disconnected, a family owned shop in which people deal directly rather than through procedure, a friend through which support is extended, a river which flows and courses with change, a series of semi-isolated villages with intermittent communication between villages at the whim of the local chiefs, and a lava lamp which can move and change in all directions and at different speeds. In a culture with these characteristics communication is a cohesive factor because it is a way to know what is going on.

D'Aprix (1999) asserts that the main dynamic of culture is the leadership. D'Aprix sees ineffective communication processes as a product of the beliefs of an organisation's leadership – mirrored in the communication system and behaviour of the organisation. Although the culture of the organisation in which the project was undertaken is a networked organisation, the potential for full staff participation is somewhat limited, due to the owner's particular leadership style.

The forum was an attempt to foster collaboration and staff participation. While collaboration does happen without the forum, there are still a number of staff in the organisation that do not feel listened to. The lack of participation has not changed over the intranet development and implementation period. The statement has been made that participation in the culture is extended on condition that "assertive initiative" is shown. This means that many in the organisation either are not listened to or do not believe that their participation will make any difference. On the other hand, for those that show assertive initiative, the opposite is true. They have various ways of involving themselves in change within the organisation and it is a mute point whether the intranet would make any difference to them – they can communicate and achieve change and difference in the organisation without any need for the Intranet.

However, feedback from the interviews indicated that the intranet has helped the senior management to communicate more with staff about what is going on. The intranet has had a positive effect on how things are done, particularly in regard to communication.

Thus, the assertion that the process of developing and implementing a user-centred intranet will "open up" an organisation's culture has not been clearly proven. The intranet will have some affect because it will encourage a higher level of communication from the leadership and enable a higher level of communication from all staff. The comments from staff show that the organisation's culture is dominated not by an intranet but by people. Even though my role in the organisation provides the opportunity to train staff to be more participative, the owner of the organisation primarily sets the cultural tone, as he holds the majority of the power. My involvement with management interventions in developing and implementing a user-centred intranet may have influenced the culture to be more collaborative but not to be significantly more participatory. Therefore it would be more realistic to conclude that a user-centred intranet 'influences' an organisation's culture, rather than opens it up.

The importance of the role of a communications manager is reiterated as far as cultural change goes. A communications manager can influence the leadership of an organisation working to link channels of communication in with organisational goals. So for example, if he leadership want to improve sales on any particular product, staff involved in the production or provision of that product can be encouraged to collaborate for improvement ideas. However, the leadership of an organisation need to embrace the networked organisational model in order to encourage such activity. Only then can dynamic cultural change be expected to occur.

3.5 The effect of a user-centred intranet on knowledge management

"Soft" knowledge or "know-how" can effectively be captured in a forum. As discussed in the literature review, Malhotra (2000) implies a knowledge management approach where the organization's members define problems for themselves and generate their own solutions. For this to be achieved through an intranet a forum would need to realise its potential. As discussed above this would be a too optimistic approach governing the organisational culture where the project was undertaken.

"Hard knowledge" refers to that which can be recorded and used for manuals or recorded in some way as intellectual capital. I created an auto-updating menu system for the intranet which accessed the manual files from the shared network drive. However, despite my best efforts I could not get this to work cohesively on the intranet. Some staff would not follow the guidelines for creating, updating and storing files on the intranet, and continued to work from the shared network drive.

I was not able to prove the robustness of a folder ordering system on the organisation's network. This was partly because I attempted to get department managers or assigned staff to update their own manuals. In this process only some people followed the guidelines, while others would not. Only those staff who were administratively strong followed the guidelines. If the manuals had been centrally controlled I may have succeeded, although that would not be true to a networked paradigm. Furthermore, our organisation would not provide the resources to the communication manager to achieve this. If an organisation was focussed to achieve creation of manuals by the staff doing the work, I believe there would need to be a lot of coaching to maintain consistency. Consistency is possible when one person is responsible for any job, as long as there is cooperation from staff involved. It is another thing entirely to get a group of people working to a standard without a large amount of supervision and training. When we are talking about a networked organisation we are talking about one where staff are contributing in a creative way. I cannot se how this is realistic.

Developing and implementing a user-centred intranet will open up an organisation's knowledge management in a participatory way, only on the condition that the filing structure is controlled according to guidelines and that it is robust enough when there is a high rate of change in the organisation. I believe that hard knowledge simply has to be centrally supervised in order to maintain consistency and the under resourcing for this at Romus Colleges is a fault of the organisation.

In discussing the issue of whether hard knowledge should be centralised, Nielsen (2002) focuses on retaining the centralized information sharing potential of intranets. He points out that intranets are still just an internal network of information. The auto-updating menu interface I created for the intranet was not used as there seemed no advantage to users compared with the commonly used windows interface to the shared drive. For an organisation in one location, the benefit of using an intranet for this purpose is negligible. If however, an organisation is geographically disbursed the idea of centralising hard knowledge management via an intranet makes sense, but would require resourcing for centralised control of hard knowledge to achieve it.

I have said that the guidelines for hard knowledge management need to be robust enough for change. The reporting mechanisms I designed for one department to use on the intranet worked for a time, but with the fast pace of change in the organisation, there was a restructuring of the reporting system along collaborative lines. This new system was a face-to-face event, which far outweighed the benefits of reporting through an intranet interface.

The process of developing and implementing the user-centred intranet at Romus has not opened up the knowledge management directly, but has had a spin-off benefit. The process led to an improved reporting system and a restructuring of the manual filing system and guidelines. These changes were not to be a part of the intranet, but they were a great improvement from previously when the manuals had fallen almost completely out of use and reporting was sporadic within a chaotic culture. Many duplicate and irrelevant manual documents were culled from the system, a restructuring of the manuals and department folders on the shared drive was achieved, the communications manager took over the security access to the folders in cooperation with IT, and an internal audit requiring department managers to account for the state of department related manuals and requiring action plans for their developments was undertaken. A resorting of HR and management documents was achieved and these documents are now made available for all staff access.

Another spin-off benefit that occurred was cooperation between the IT and the communication manager. I took over the responsibility for setting security access for folders on the shared network drive. This came about because when a new staff member joins the organisation they need to be allocated to a security group, which in turn gives them certain permissions for accessing folders on the shared drive. Since my office managed human resources, we were most directly aware when staff roles were being adjusted and which areas new staff should have access to.

3.6 Contribution of this project to knowledge

The ideas I wanted to explore have been done so in a robust environment, where my role was that of a key manager. Significant leeway was possible with the degree of experimentation and management interventions were integrated with restructuring activities and the consequential re-jigging of related systems, such as administration, reporting and communications. During the process there was a wrestling with the issues of an organisation operating within the environment of a knowledge based and global economy and a consequential embracing of a networked paradigm.

The role of communication management surfaced as a requirement for managing a user-centred intranet and its coexistent technological applications within the context of managing communication channels. This finding is a contribution to the general field of knowledge on networked organisations. It has contributed to the role definition of communication managers and shown a number of boundary crossing activities that a communications manager would likely be involved in.

There is little material available on this subject and no material with an insider's point of view so explicitly opening up the subject. The most applicable material came from consultants, who specialised in either intranets and/or communications, where they also enunciated the values of networked organisations. The findings of this project will be most pertinent for those working in these professional areas - the fields of intranet and/or communications consultancy. Managers involved in restructuring organisations and driving internal change (metastructuring) towards networked organisations will also gain insight from the findings. Leadership can better understand the strategic importance of user-centred intranets and communication management if they wish to create a collaborative and participatory culture in today's modern business world.

For leadership and management the spin-off benefits have been highlighted. The value can be seen for resourcing a review process to redevelop and re-implement "user-centred" features into existing intranets. It provides a focus, for change for an organisation evolving into a networked organisation, for a communication manager to undertake. The process of developing and implementing the user-centred intranet at Romus Colleges was very much an action research project which evolved through development, testing and feedback cycles.

As DeSanctis & Fulk (1999) comments on the dynamic nature of this process which "support episodes of experimentation, reflection, and change in technologies and their use, so as to allow for the evolution of technological frames, work habits, and communication routines in conditions of change. (P. 133)

The networked organisation is a paradigm shift from the more hierarchical approach. Everything changes. Among the characteristics are that there is more merging of the boundaries of work groups and departments, greater collaboration and communication networking, responsibility shifting to staff to take initiative for outcomes with flexibility to design their own procedures. Leadership is a key requirement for this paradigm shift to come about. An intranet could facilitate such a change, but the core culture of the organisation, which is mainly created by the leadership of an organisation, is considered the primary focus if an effective networked organisational model is to evolve. This has been reinforced by this thesis. But the project has shown that developing and implementing a user-centred intranet has brought the networked organisational model to the fore. Since this model is an effective model for today's business success, introducing or adapting intranets with a user-centred focus is necessary.

Sviokla (2004, p. 1) suggests that technological uptake plays a significant role in "bringing new business structures to the fore" and that since "management structures ultimately make the difference between a company that adjusts to the environment of the 21st century and those left behind" technological development and implementation has a strategic role.

3.7 Conclusion

The first propositions is affirmed. A user-centred emphasis in intranet design is necessary in today's knowledge based global economy, where there is a high rate of change. In addition, the networked organisational model should be embraced as a flexible model catering for change by promoting organic and dynamic networking within an organisation. This environment is communication rich and collaborative. A user-centred intranet is a communication channel well suited to support this model.

The second proposition is conditionally affirmed. A user-centred intranet will not only act as an artefact, but also the process of developing and implementing it will open up an organisation's culture, communication, and knowledge management in a participatory way. However, the degree that this will be achieved depends on a number of factors. The leaders of an organisation need to embrace a networked organisational model, encouraging and resourcing communication management for a culture of participation and collaboration.

References

Agar, M. H. (1986). *Speaking of ethnography.* Newbury Park: Sage.

Anderson, C. (1995, 1 July). The accidental superhighway. *The Economist,* pp. 1-26.

Babbie, E. (1983). *The practice of social research* (3rd ed.). Belmont, CA: Wadsworth.

Becker, H. S. (1958). Problems of inference and proof in participant observation. *American Sociological Review, 23,* 652-60.

Boyce, R. (1997). *The communications revolution at work: The social, economic and political impact of technological change.* McGill: Queens University Press.

Bolter, J. D. (1984). *Western culture and the computer age.* Chapel Hill: University of North Carolina Press.

Becker, A. L. (1982). *Field work evidence in sociological work: Method and substance.* New Brunswick, NJ: Transaction Books.

Carlston, D. *(1996).* Digerati: Encounters with the cyber elite. *San Francisco: Hardwired.*

Castells, M. *(1998).* The rise of the network society. *Oxford: Blackwell.*

D'Aprix, R. *(1999).* Communication in Effective Organizations. In A. Wann. (Ed.), Inside organisational communication *(pp. 1-10) (3rd ed.). IABC. NY: Forbes.*

DeSanctis, G., & Fulk, J, (Eds.). (1999). *Shaping organization form: Communication, connection and community.* Thousand Oaks: Sage.

Doyle, J. (2002). *New community or new slavery: The emotional division of labour.* London: The Industrial Society.

Feenburg, A. (1996). *Heidegger, Habermas, and the essence of technology* [Talk at the International Institute for Advanced Study, Kyoto]. Retrieved 12/8/2003, from http://www-rohan.sdsu.edu/faculty/feenberg/kyoto.html

Foucault, M. (1988). Technologies of the self. In L. H. Martin, H. Gutman & P. H. Hutton (Eds.), *Technologies of the Self: A Seminar With Michel Foucault* (p. 166). Amherst, MA: The University of Massachusetts Press.

Gayesh, D. (2000). *Managing the communication function: Capturing mindshare for organisational performance.* San Francisco: IABC.

Gregory, K. L. (1983). Native-view paradigms: Multiple cultures and culture conflicts in organisations. *Administrative Science Quarterly, 28,* 359-376.

Hagel, J., & Armstrong, A. (1997). *Net gain: Expanding markets through virtual communities.* Boston: Harvard Business School Press.

Hagel, J., & Singer, M. (1999). *Net worth: Shaping markets when customers make the rules.* Boston: Harvard Business School Press.

Harding, R. (2002). *Re-inventing the wheel? Productivity, performance and people.* Work and Enterprise: The Work Foundation.

Habermas, J. (1994). *Structural transformation of the public sphere.* Chicago: MIT Press.

Heidegger, M. (1977). *The question concerning technology.* (W. Lovitt, trans.). New York: Harper and Row. (Original work published 1949)

Heltz, S. (2002). *The intranet advantage: Your guide to understanding the total intranet and the communicator's role* (2nd ed.). San Francisco: IABC.

Hesselbein, F., Goldsmith, M., & Beckard, R. (Eds.). (1997). *The organisation of the future.* San Francisco: Jossey-Bass.

Hildebrand, C. (1999). Does knowledge management equal IT? [Electronic Version] CIO Enterprise Magazine, September 15, 1999. Retrieved July 10, 2004, from http://www.cio.com/archive/enterprise/091599_ic.html

James, J. (1998). *Sex at work: A survival guide.* London: The Industrial Society.

Kouzes, J., & Posner, B. (1993). *Credibility: How leaders gain and lose it, why people demand it.* San Francisco: Jossey-Bass.

Kounalakis. M,, Banks. D., & Daus. K. (1999). *Beyond spin: The power of strategic corporate journalism.* San Fransisco: Jossey-Bass.

Lea, M., O'Shea, T., & Fung, P. (1995). Constructing the networked organisation: Changing organizational forms and electronic communications. In G. DeSanctis & J. Fulk. (Eds.), *Shaping organisational form: Communication, connection, and community* (pp.295-324). Thousand Oaks: Sage Publications.

Lipnack, J., & Stamps, J. (1993). *The teamnet factor: Bringing the power of boundary-crossing into the heart of your business.* New York: John Wiley & Sons.

Lipnack, J., & Stamps, J. (1994). Organizing principles of teamnets, with practical hints. *The age of the network: Organising principles for the 21st century.* New York: John Wiley & Sons.

McChesney, R. W. (1995). *Telecommunications, mass media and democracy.* Oxford: Oxford University Press.

Malhotra, Y. (1997). Knowledge management in inquiring organizations. *Proceedings of 3rd Americas Conference on Information Systems.* Philosophy of Information Systems Mini-track. (pp. 293-295). Retrieved July 7, 2002, from http://www.brint.com/km/km.htm

Malhotra, Y. (2002), Why Knowledge Management Systems Fail? Enablers and Constraints of Knowledge Management in Human Enterprises. In M.E.D. Koenig & T.K. Srikantaiah (Eds.), *Knowledge Management Lessons Learned: What Works and What Doesn't*, Information Today Inc. (American Society for Information Science and Technology Monograph Series), 87-112, 2004. Retrieved September 2, 2005, from http://www.brint.org/WhyKMSFail.htm

Marcuse, H. (1964). *One-dimensional man: Studies in the ideology of advanced industrial society.* Boston: Beacon Press.

McKie, A. (2002). *Virtual value: Conversations, ideas and the creative economy.* London: The Industrial Society.

Molich, R. (2001). *230 tips and tricks for better usability testing report* [Electronic version]. Retrieved June 2, 2002, from the Neilson Norman Group website: http://www.nngroup.com/reports/tips/usertest/

Neilson Norman Group. (2001). *Intranet Design Annual: 10 best intranets of 2001* [Electronic version]. Retrieved September 5, 2002, from http://www.nngroup.com/reports/intranet/2001/

Neilson, J. (2002a, December 3, 2001). *Jakob Nielsen on usability and Intranets* [Interview]. Retrieved November 12, 2002, from http://www.it-director.com/article.php?id=2383

Nohria, N., & Eccles, R. (1996). Face-to-face: making network organizations work. In N. Nohria & R. Eccles (Eds.), *Networks and organisations: Structure, form and action* (pp. 288-308). Boston: Harvard Business School Press.

Negroponte, N. (1995). *Being digital*. Hodder Stoughton.

Schein, E. H. (1993). *Organizational culture and leadership* (2nd ed.). San Francisco: Jossey-Bass.

Santosis, M., & Surmacz, J. (2001). *The ABCs of knowledge management* [Article]. Retrieved November 20, 2002, from http://www.cio.com/research/knowledge/edit/kmabcs.html

Schwartzman, H. B. (1993). *Ethnography in organisations*. Newbury Park: Sage.

Skyrme, D. (1999). The networked organization [Article]. Retrieved August 17, 2002, from http://www.skyrme.com/insights/1netorg.htm

Turkle, S. (1995). *Life on the Screen: Identity in the Age of the Internet.* New York: Simon & Schuster.

Turkle, S. (1998). The Cyberanalyst. *Digerati: Encounters with the Cyber Elite.* Edited by John Brockman. San Francisco: Hardwired.

Holetin, R. (1997). *Composing cyberspace: Identity, community and knowledge in the electronic age.* McGraw Hill.

Virtanen, M. (1997). *The role of different theories in explaining entrepreneurship.* Proceedings of the 1997 USASBE Conference. This reading can be found at the Helsinki School of Economics and Business Administration website: http://www.usasbe.org/knowledge/proceedings/1997/P109Virtanen.PDF

Wann, A, (1999). *Foreword: Inside organizational communication* (3rd ed.). IABC. NY: Forbes

Winstone, B. (1998). *Media technology and society, a history: From the telegraph to the internet.* London: Routledge.

Yin, R. K. (1994). *Case study research: Design and methods* (2nd ed.). Thousand Oaks: Sage.

Index

Access to the intranet, 120
Book structure, 11
Communication management, 45, 61
Communications, management and culture, 53
Communication success, 118
Communication systems of Romus, 105
Cultural features of the organisation, 17
Culture of Romus, 103
Feelings of belonging, 46
Knowledge management, 66
Knowledge management ideas, 69
Knowledge management of Romus, 108
Influence of the wider business context, 101
Interview feedback, 85
Introduction to the organisation, 12
Introduction to theory, 29
IT and New Media departments of the organisation, 19, 20
Management personnel in the organisation, 14
Move towards user-centred intranets, 9
My role in the culture of the organisation, 18
Networked organisations, 115
Overview of the organisation, 13
Owner and president of the organisation, 15
Potential of network digital technologies, 49, 58
Proposal, 22
Relationship management, 70
Reporting, 29
Responding to users, 34
Rolling out the Intranet, 49
Socio-political theories of technology, 40
Social necessities and key benefits, 41, 43, 44, 45, 47
Stages in the project development, 13
Technical emphasis, 35, 50, 55, 121
Technical issues, 116
Technology used, 20-21
Training and maintaining, 121
Training sessions and intranet uptake, 54
Warnings from the Philosophy of Technology, 31, 37, 38
What I wanted to explore, 6
Why I took on this project, 7

Printed in the United Kingdom
by Lightning Source UK Ltd.
134348UK00001B/179/P